Mathematical Ecology

Mathematical Ecology

edited by
Peter Allen (Brussels), Werner Ebeling (Berlin),
Manfred Peschel (Berlin), Peter Schuster (Vienna),
Yuri M. Svirezhev (Moscow)

Mathematical Ecology deals with mathematical models of evolution processes, in the biosphere as a unity of growth and structure-building, with spring-up of new species, their interaction and possible extinction, with the impacts of human activities on the environment and the corresponding consequences for the biosphere.

The whole field seems to be a certain amalgam of suitable system philosophy, in which real world phenomena are considered through an ecological pair of spectacles with the help of system methodology and mathematics.

All publications published herein are of interdisciplinary interest for ecology, biology, economy and technical engineering.

Applied Simulation of Marine Communities with Reference to the Fouling Community

by Mikhail Vladimirovich Gal'perin

With 48 Figures and 29 Tables

Akademie-Verlag Berlin

Author

Dr. Mikhail Vladimirovich Gal'perin
State Committee on Standards
of the USSR Council of Ministers

Reproduction of the original author's manuscript

ISBN 3-05-500761-1

Erschienen im Akademie-Verlag Berlin, Leipziger Straße 3—4, Berlin, DDR-1086

Lizenznummer: 202 · 100/90
Printed in the German Democratic Republic
Gesamtherstellung: Druckerei „Thomas Müntzer", Bad Langensalza, DDR-5820
Lektor: Dipl.-Phys. Gisela Lagowitz
LSV 1075
Bestellnummer: 764 076 9 (9278)

Foreword

Mathematical Ecology as a scientific discipline took shape as early as in the beginning of this century, but it was not until the 1950s and 1960s that ecological research covered important objects and yielded important scientific findings. Aquatic ecology is a major field of ecological research. Its objects of study are complicated and hard to access, the research is costly and erroneous forecasts and estimates of ecological changes may do much harm. The pollution of rivers lakes, and shelf areas in the World Ocean becomes dangerous and has a significant impact on the development of civilization and economics. Mathematical modelling is designed to reduce the cost of ecological research to improve the forecasting accuracy, and to yield new findings about the life of aquatic ecosystems. The dramatic scaling of the human activities in the shelf areas, internal, and open seas calls for the development of research and forecasting models, especially of bottom ecosystems. It has been discovered that bottom communities have a profound impact on the status of plankton systems and strongly influence the productivity of shelf waters and the monitoring potential. The advent of new generations of computers has resulted in numerous simulation models of aquatic ecosystems. These models do not only yield phase portraits of the system, they also provide an insight into the dynamics of both the components and the entire system.

This book describes the basic applied aspects in the simulation of marine ecosystems, more specifically, the fouling community. The model developed by the author and tested over years has the best features of applied ecology. Tenacious analysis of the field data, its statistical processing, made the model biologically sound in that the mathematical formalization tools as well as the computer implementation of the simlation program are consistent with the biological version of the object. This "homage" to the object's ecology and the biological laws of its development has, unlike a welter of theoretical and semi-theoretical papers on mathematical ecology, borne tangible ecological fruit. Firstly, both the unified applied approach to the development of a model and the model itself proved ecologically meaningful. Secondly, the author did not have to twist the numerical findings of simulation to make them fit the ecological concept. Thirdly, the implementation of the simulation model of the fouling community as a typical bottom community led to the formulation of new ecological laws of development and life of the community, and tested the applied methods developed by the author. Fourthly, for the first time in marine ecology, the simulation experiment yielded numerical estimates of the rates of intra- and interspecific (including topic) relations between the populations belonging to that community. Fifthly, the use of standard ways to build mathematical models of an aquatic ecosystem is conducive to new ideas in the procedures of developing simulation models of marine ecosystems and to the description of the principles of the event approach to their design and development.

The author's thoroughness in presenting the material is an undisputable asset of the book as well as of the numerous Figures.

The book will undoubtedly be useful and significant for applied biologists, mathematical ecologists, engineers, and researchers in the development and implementation of models, and for undergraduate and graduate students in this field.

M. Ye. Vinogradov,
Corresponding Member, USSR Academy of Sciences
Professor of Hydrobiology

For my parents, Vladimir Gal'perin
and Sofia Margolina

Preface

Numerous responses to the basic results reported in my earlier articles on the simulation of communities, in particular of the bottom community, convinced me that all stages in simulating both the entire bottom community and its components must be described in detail in one book. The research was under way from 1976 to 1985. That period was long enough for some approaches to programmed run of simulation models to change. Thus, personal computers were taken over from IBM 360/370. I tried to omit whatever could not keep pace with the technology. The energy approach was used. Considering the development of the community and the functions of populations in terms of energy metabolism made it possible to assess the energy flows through bottom communities for the first time. I made a conscious effort to provide mathematical tools as straightforward as possible, firstly, because in combination with energy budget equations this approach paved the way to an assembly of the community models from models of most primitive energy processes and, secondly, to make the mathematical formalization of the population and community models comprehensible to the ecologists. For the same purposes, an independent notation was used in mathematical formalization of the models in every section. I gave priority to the ecology of the objects of modelling and to the ecological interpretation of the numerical results, especially because they are largely my own. I am proud of the numerical estimation in energy and weight units of the rates of inter- and intraspecific relations in the communities. I also believe that some new ideas in computer simulation in ecology, which evolved during the development of simulation models, will be offered. I endeavoured to make the book interesting for scientists and engineers of various applied fields. The final judgement of this attempt is of course, up to the reader.

I take this opportunity to express my gratitude to
- my parents, who made it possible for me to concentrate on this research for ten years;
- M. Ye. Vinogradov, Corresponding Member, USSR Academy of Sciences, and Professor Yu. M. Svirezhev for their support and advice;
- The Professors S. A. Levin and B. S. Patten for making available numerous useful papers.

Dr. Michael V. Gal'perin

Contents

INTRODUCTION

The environmental problems have grown beyond national boundaries today. Moreover, the preservation of the biosphere as an international priority is becoming a moral and political as well as scientific issue. As a result, research in this field becomes increasingly relevant. The dramatic sophistication of economic systems calls for ever growing investments of research efforts in the analysis of the ecological consequences of human activities in both dry land and the marine environment. The development of ecosystems has to be reliably forecasted if nature conservation decisions are to be made intelligently. Consequently, methods resulting in credible ecological forecasting are urgently needed. One of such methods is the simulation, which emerged several decades ago as a tool for detailed studies of the behaviour of complex systems in qualitative and quantitative terms. The high cost of field data and the complexity of causal relations in ecological systems has hindered a breaking forth in the simulation of ecosystems. For this reason, the expertise in control at lower levels of the ecosystem and community may shed light on what is to be done on the global level. The description of formalized methods yielding tangible ecological results may be equally useful.

The environment needs protection, in particular, when construction is under way on the coasts and shelves. Water intake, drilling, and port installations have hard surfaces which, unless protected, are populated by marine organisms. The resulting fouling community is a specific component of the bottom ecosystem in a seaarea. In natural conditions, e.g. on rocks, as well as in underwater parts of various man-made structures the fouling consists mainly of attached species, among which mobile forms develop, too. This community increases the roughness of surfaces, leads to the corrosion of metals, reduces the diameter of water intake pipes and the speed of ships. For this reason biooceanology has to know more about the fouling communities in order to develop ways to reduce the economic losses. Biological control of fouling is thus an integral part of biosystem control and an intelligent utilization of natural resources. To combat fouling, the rates of development and production processes have to be estimated both for their members and the entire community. Fouling communities are highly productive systems. They consist of a few species and the stability of the community is relatively low. Such communities for themselves are good models for studying the functioning of bottom communities consisting of a few species, their production rates, interspecific relations, and the impact of abiotic factors. Therefore, research in the functioning of the fouling community and the energy flows through them is useful not only economically but also scientifically (Starostin and Permitin, 1963; Dolgopol'skaya, and Brayko, 1974).

Topic relations resulting from the competition for space are more important than nutritional relations in numerous bottom communities. The dynamics and quantitative estimates of the structure and functioning of these relations are practically unexplored because the criteria for their quantitative estimation are unclear and the objects are not easy to study. Fouling communities are, however, the most accessible objects of this kind. Therefore, the development of methods for evaluating the functioning of fouling communities is one stage in studying the entire range of bottom communities.

Bottom communities are not readily accessible (Nesis, 1977). On the other hand, data on some quantities that dictate the system functioning would be useful in establishing a workable model. Since field experimentation with marine ecosystems is undesirable in most situations, modelling is an indispensible tool. There is a consensus that, in ecology, models need not copy the actual system (Watt, 1968; Odum, 1971). In most cases the model is a reasonable simplification of the natural object in which, however, the basic processes are reflected and the system functioning can be investigated. Studies of basic processes which dictate the biosystem development have relied on mathematical techniques for the description and explanation of natural phenomena since the last century when Verhulst and Mobius published. But mathematical ecology gained momentum very soon, namely, in the early 20th century when Volterra (1926) Lotka (1925), and Gause (1933) reported their findings. These researchers tried to estimate quantitatively and to describe interspecific relations such as predator - prey and parasite - host. The analytical solution of the mathematical models they proposed revealed certain new patterns in the evolution of coexisting populations. The sophistication of Volterra's ideas culminated in a new kind of modelling and development of a mathematical theory of survival. Research along these lines is still pursued by our contemporaries (for example, in Soviet science, Gimel'farb et al., 1974; Svirezhev and Logofet, 1978).

In marine ecology the modelling was persued by Sverdrup, Fleming, Cushing, Riley, Steele, and,in the USSR,by Poletayev (1966), Lyapunov (1971); Menshutkin, (1971), and Vinogradov et al. (1971, 1973). Until the computer boom in the middle of the 1960s, models in this field were intended for analytical research (Fleming, 1939; Riley et al.,1949; Cushing, 1953, 1959; Svedrup, 1953; Kierstead and Slobodkin, 1953). The first computer programs developed in the late 1940s and early 1950s were mainly procedures of numerical solutions. However, already in 1949 Riley, Stomel, and Bampas, and, later, Steele and Menzel (1962) tried to develop a mathematical method of the production cycle of algae, and to perform a computer experiment. The results matched the actual data fairly well. The sophistication of computer technology resulted in principles of a new kind of research which required computer experimentation with the mathematical model. Not only the descriptive function could be implemented, but the computer could also model the behaviour of the system. This method of modelling is referred to as simulation. Therefore, the research of Riley and Steele may be considered as the first stages of simulation in marine ecology. In the 1960s and '70s, simulation became widely used in both Soviet (Vinberg and Anisimov, 1966; Menshutkin, 1971; Vinogradov et al., 1973; Gorstko and Surkov, 1976, and others) and foreign research (Parker, 1967; Patten, 1965; Eberhard and Hanson, 1969; Patten et al., 1976; Goodall, 1973). Gradually, its approach became universally accepted. First, the biological knowledge and assumptions on the system's kinetics lead to a system of equations describing macroelements of the system, and then coefficients of the mathematical models are obtained from actual data with the aid of computer programs. By varying the coefficients and computing options, the desired situation is simulated.

When the model is designed to obtain detailed knowledge and description of the object, quantitative values of the coefficients can nor be found from every case. Writing a mathematical model of the entire system may prove difficult without staged studies of the structure and functioning of the components, notably the energy flows in communities, especially bottom communities, because without this step the coefficients of transfer functions (energy flow rates from component to component) cannot be estimated. For this reason, the investigation of the energy of bottom communities by models usually starts with a mathematical description of the object in terms of input - output equations. This has to be done in several stages because of its high complexity (Vinberg, 1962; Lyapunov, 1974; Vinogradov and Menshutkin, 1971; Whittaker, 1975; Odum, 1971; Odum and Odum, 1976). Once the composition of the system is clear, the energy flows in the components and relations between them are studied. Proceeding in such a way the tasks for the model can be formulated.

The success in the development of simulation models depends largely on the significance and amount of field data, as well as on the soundness of the biological theory of system development. In order to have standard approaches for the development of workable mathematical models leading to simulating programs, the object of research should most probably be an ecosystem accessible for direct observation, from which biologically interpretable simulation results can be obtained. Simulation is, in my opinion, only successful if the model is biologically sound. Good results have been recently obtained by combining sound biological theory and efficient, although fairly easily applicable, mathematical techniques. On the other hand, strong mathematical tools do not work with unsoundly applied biology.

The bottom fouling community combines typical features of the bottom marine community as well as of a system subjected to economic and technical factors. Besides, the fouling community in Zhdanov on the coast of the Bay of Taganrog in the Sea of Azov does a lot of damage to the water supply facilities of the metal works (Starostin, 1963; Starostin and Permitin, 1963). It is also important to know and estimate quantitatively the rates of various energy transfer processes in the community, in particular, the interspecific relations. As noted above, the fouling community is a convenient natural model of the bottom ecosystem, whose study will be a significant contribution to biology of the entire bottom fauna.

Biological research investigated the composition of the fouling, the basic features of most widespread species, and the dynamics of settlement of their larvae. The need to forecast the development of the fouling community and ways of biological control with the aim to reduce the costs for the metal works called for a mathematical model of the processes in the formation and functioning of the entire community and its components. This model reflects the knowledge about the community ecology and observation data. The latter were obtained in experimental 1 m^2 plates on which the main populations were observed during several spring - summer periods, both individually and in interaction. Energy flows of individuals of every main species in the community were obtained. The data on the nutrition were represented as mean concentrations of basic sources, phytoplankton and zooplankton, in one cubic meter of water. Field experimentation supplied data on the impact of basic abiotic factors.

The analysis of the research objectives, requirements to the model, and the kind of field data led to the formulation of the goal for the development and computer implementation of a simulation model of the fouling community and for the carrying out of a predesigned simulation experiment. Once the goal is formulated, the research stages can be identified.

At the very beginning it is necessary to introduce and stringently define the basic concepts. A detailed description of the simulation object will make it possible to discuss the main features of the approach used in the model development and to prove the legitimacy of the way in which the simulation model was developed and implemented. This stage will culminate in the formulation of specific research goals.

At the second stage, appropriate models will provide an insight into the most important processes in the evolution and functioning of the most widespread species of the fouling community, which develop in isolation from each other with the aim to estimate quantitatively and understand better the vital functions and energy flows in them. Population monocultures will be described in Chapter 2, which will also provide statistical estimates of the main empirical relations used in the models.

The system of interspecific biocoenotic relations, the method of quantitative estimation of their rates, the mathematical formalization of the processes of the interaction of species and their computer implementation as the core of the simulation model will be described at the third stage, which will also be devoted to a predesigned simulation experiment, the forecasting potential of the model, and sensitivity to variations of the main parameters. At the end of that stage an attempt will be made to predict further lines of research in computer-aided simulation and setting up a special-purpose simulation system.

Still another stage will be the description and discussion of an essentially new approach to simulation in marine ecology, in which the ecological object is described in terms of events, and therefore SIMULA, PROLOG, LISP, and other modelling languages are used.

Statistical processing of the simulation results is important at the final stage, for it provides quantitative data for the quantitative description of the interaction between populations of the fouling community and leads to the formulation of important patterns of species interaction in bottom communities.

Chapter 1. THEORETICAL FUNDAMENTALS OF SIMULATING MARINE ECOSYSTEMS

1.1. Basic Definitions

An analysis of the way in which complex multi-component systems operate and of the control of such systems is never easy because of the very intricate interactions between the components. In particular, the behaviour of ecosystems in response to changes of exogenous factors is unpredictable in general. Since such systems are hard to access and occasionally impossible to experiment with, modelling has been extensively used as a research tool. Modelling is normally understood as investigating some objects indirectly through the analysis of other objects. In this light, modelling can be classified as physical, analogous, and symbolic. This research will concentrate on the latter feature, which includes mathematical modelling, in particular, simulation (Buslenko, 1978).

Shannan (1975) defined a model as representing an object system or idea in a form different from their actual existence in order to help to explain, understand or improve the system. This definition logically follows from the above understanding of modelling. In most cases, modelling, notably symbolic, is applied to complex objects or systems. The concept of a complex system has been widely discussed in the scientific literature (Good and McCall, 1969; Kalman et al., 1969; Buslenko, 1978; Zeigler, 1976, 1984). Since this research is not devoted to purely theoretical issues in modelling, a fairly general and widely used definition of a complex system will serve its surposes. A complex system is a multi-level structure of interacting elements making subsystems at different levels. The elements cannot be further decomposed at the lowest level (Buslenko, 1978; Buslenko, 1978; Zeigler, 1984 a & b). The choice of model elements is dictated by the objectives of model research. In most mathematically ecological research projects the elementary levels are associated with levels in the organization of the living nature. Timofeev-Resovskiy (1964) spoke of the cellular, organismic population, and biogeocoenotic levels. This approach to define the elementary level of a mathematical model has been widely used (Watt, 1968; Williamson, 1972; Whittaker, 1975; Parker, 1973) abroad and in the USSR (Menshutkin, 1971; Vinogradov and Menshutkin, 1977; Voronich et al., 1981). The system structure, or hierarchy, is usually reflected in the morphological description, Fig. 1. This is supplemented by the description of data flows. The functional description, which is given most attention in ecological papers, reflects changes of the system state within time, under the impact of endogenous and exogenous factors (Patten et al., 1976; Menshutkin, 1971; Belyaev, 1980) as transition statements (controls, algorithms, etc.) which depend on the set of times, system states, and processing leading to an impact on the system. Usually, a morphological description precedes the stage of notating a functional description. The most important step in obtaining it is the structuring of the complex system. The structuring is a decomposition of the complex system into subsystems and elements and their representation as a multi-level hierarchical structure in the form of a structural flowchart. The structuring becomes largely an art, notably in ecology where the dominating factor is the biologist's view of the system. Structural flowcharts have not been, to the best of my knowledge, the subject of any paper. On most occasions, systems are represented as a functional flowchart made of actual ecosystem components and links between them. In the associated mathematical models, the structuring process is confined to recording the right-hand sides of systems of equations. On the other hand, in papers on theoretical issues of modelling, the structural flowcharts and the interfacing of elements are usually more abstract because of the requirements of mathematical formalization and software implementation of a complex hierarchical element interaction and relation system (Good and McCall, 1969; Naylor, 1971; Shannon, 1975; Buslenko, 1978; Zeigler, 1984 a & b). For this reason, if detailed models of the object suitable for extensive computer experimentation are to yield a forecast of the system behaviour in various functioning conditions, special attention must be paid to the development of structural flowcharts and their software implementation.

Most of the complex systems that have been studied so fare are dynamical systems in the sense that they change their states in time response to exogenous and endogenous factors. In biological systems the structure of inter-component relations is also dynamical. As a result, the mathematical tools become more unwieldy and difficult to realize in the software. It was not until the computing technology became very powerful that a compromise could be concluded between the number of necessary numerical operations and the time for their execution. Larger memories and faster computers made more expressive programming languages possible. Application program packages and libraries of algorithms became major auxiliary tools. For this reason, algorithmical descriptions or algorithmical formalization make the core of up-to-date multi-program systems. Algorithms relying on a significant amount of expert heuristic information and, thus, are more than a strict sequence of activities, are referred to as logical-heuristic algorithms. In ecology they are extensively used, notably where game mechanisms are involved (Bear, 1972; Bukatova, 1979; Kontorin, 1980; Jansson, 1972, 1978; Patten, 1976; Zeigler, 1984 b).

In the first attempts using computers in modelling natural processes only the speed of hitherto manual operations was increased. It was not before comprehensive studies of large-scale complex systems were

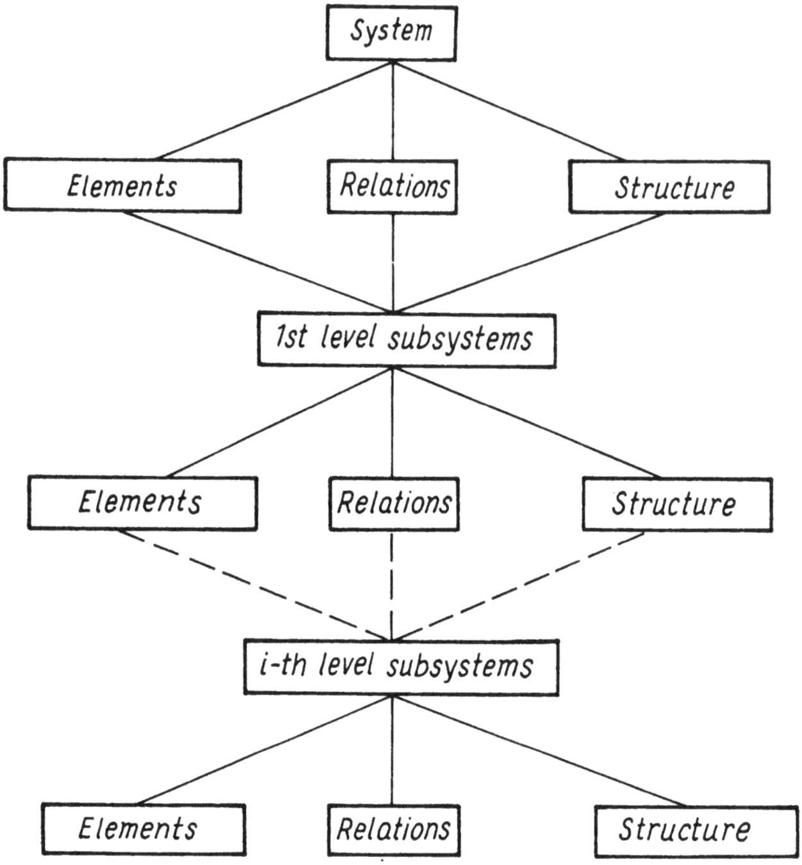

Fig. 1 Hierarchy of morphological description

made that the principles of systems analysis and its instrument, simulation, were developed.

1.2. Simulation and Its Application in Marine Ecology

Simulation as a concept is still in its formative stage; in most cases it is understood as a numerical computer experimentation with the mathematical model of a complex dynamic system. In other words, simulation is the computer representation of a general system situation and a sort of replay of phenomena and events of the process, rather than experimenting with the actual system. A simulation model is complex software that formally describes the object in mathematical terms and enables the simulation. Simulation provides the description, explanation, and forecasting of the behaviour of the system under different conditions.

Initially, a simulation model was a narrower concept than it is now. Such models were once the plainest possible computer programs that differed from the standard numerical procedures of equation solving only in the fact that the program was supplemented with several blocks executing an additional "game" algorithm. Then, simulation models proper followed in the aerospace industry. As this modelling technique found applications in economy and stock exchange, theoretical papers tended to define simulation and its principles in more stringent terms. However, simulation remains a method applied practically to any model which is in any way related with computer experimentation, e.g. by the choice of coefficients for systems of equation and computation of various paths for solutions of these systems of equations. Therefore, to provide a better definition of the simulation model and simulation as a process, let us have a closer look at the guidelines of systems analysis.

Basic Simulation Principles

The first computers used in modelling had a small core memory and were slow. Since it requires to process a fairly large amount of data, the development of simulation depends largely on the hardware potential. With the advent of the third- and fourth-generation computers, whose advantages are multiprogramming operation and independent data transmission between various memories, and of personal computers, highly complex models became essentially feasible. However, as the complexity of the system to be modelled increases, so does the effort to be invested in the formal description of the elements and their interaction and in programming the algorithms. Ecology faces equal difficulties in treating natural systems as entities. A major principle in simulation, is, as noted above, the structuring of the system. Unlike representing the ecosystem as several components connected by actual links (functional structure), the structural flowchart obtained by considering the same ecosystem in terms of the systems theory and simulation may be termed a system, and the process of obtaining it systems structuring. It would be important to define the interaction of elements in a structured system. Buslenko (1977, p. 24) defines it as a "joint functioning of elements whereby the behaviour or properties of one is in a general case dependent on conditions defined by properties of other elements". Most papers on the simulation of natural systems imply the actual processes by links of elements. In mathematical descriptions of these models the interactions of elements were expressed as an addition of appropriate terms in the right-hand side of the systems of equations. In contrast, the sciences which deal with the theory of simulation consider links of elements and subsystems in terms of an exchange of signals in compliance with the interfacing of elements which is dictated by the structural flowchart. The actual interaction processes in the system are represented as in dedicated models vital relations. This approach is very important also in the software implementation of the model. Therefore, the second principle of simulation is to consider the interaction of elements to be a signal exchange mechanism. The third principle is the development of a model as a feedback system in which it is possible and necessary to amend the structure and its mathematical description in response to the results of the simulation experiment, which is predesigned to serve the research objective. This principle is made possible by one advantage of simulation, namely, that in the experimentation the actual system and its model can change. For hard-to-access objects of study such as ecosystems and communities, this advantage is often decisive for the choice of the research technique. Another advantage making this principle feasible is the possibility to study communities with different time steps of the model, which is sometimes essential in order to preserve the ligitimacy of biological assumptions. With the detailed treatment of the object, as required by simulation, the well-known laws and patterns in the description of the model have to be supplemented by expert evaluations and data obtained by a direct observation of the system. In some cases there is only an algorithmic way to formalize the knowledge of natural phenomena. Algorithmic descriptions incorporate large amounts of information of different types and empirical relations for subsystems of different levels. In some papers on the simulation of complex systems in which the processes are dynamically structured such logical-heuristic algorithms provide complete formal descriptions of systems (Bear, 1972; Buslenko, 1977; Patten, 1976; Gal'perin, 1983 a-e, 1986 a & b). While in the development of mathematical models intended for analytical studies the model structure has to be simplified by aggregating the components and processes, the models embodying the principles mentioned above maintain a sophistication which is commensurable with that of actual systems, only the software potential being the constraint. Consequently, the simulation program has to be modular, flexible, reliable, modifiable, and structured (Dijkstra, 1976; Jodan, 1975; Brayer, 1975; Hughes and Michtom, 1977; Boehm et al., 1978).

The process of developing a simulation model itself is a research activity which provides an insight into the system properties (Buslenko, et al., 1973; Naylor, 1971; Radford, 1972; Patten et al., 1976; Zeigler, 1984 a). Therefore, let us recall the stages in which simulation models are developed and the basic features of models.

Stages of Development and Basic Features of Simulation Models

The development of simulation models is generally considered a nine-stage process.

1. Statement of the problem and description of the object. The result dictates both the model structure and the mathematical tools.

2. Structuring, which is mainly defined by the objectives, the model development method and, in ecology, also by biological concepts in the research.

3. The choice of mathematical tools to be used in the modelling is dictated by the specifics of the structural diagrams; if these are not developed, by the structure of the processes studied and field data introduced into the model.

4. The first three stages make it possible to choose the numerical techniques of solving the resulting systems of equations and relations, and the programming language. The software implementation stage seems to consume most efforts because simulation models usually have a complex logical structure, or a dynamical system of links between elements. Various special-purpose simulation languages such as GPSS, SIMSCRIPT, SIMULA, ISL-8, ASTRA, DYNAMO, and may be others, reduce the expense and accelerate the output of results. Every such language is intended for a certain type of model, data, and objective. General-purpose languages such as FORTRAN, PL-1, Cobol, and ALGOL-60 do not impose stringent requirements on the structure of the data or the mathematical description of the system. On the other hand, special-purpose languages are different in the structure, the way time and events are represented, and in the rules of handling the data structures, but they enable an easy modification of the structure of the model itself. Indeed, the choice of the language is entirely dependent on the preceding stages. The tree of requirements to the software, Fig. 2, emphasizes the importance of the fourth stage.

5. Test simulation is carried out in order to check the compatibility of basic, housekeeping and service modules of the set of simulation programs, and the model functioning in control conditions and with field data. If the model is found to be inadequate to the natural system, a correction is made, sometimes an additional field experiment will be necessary. All adequacy analysis methods can be classified into qualitative and quantitative ones. The first mainly compare the functioning in similar conditions of the natural system and its mode, while the latter concentrate on the difference between the numerical results of simulation and benchmark data of the field experiment for similar processes.

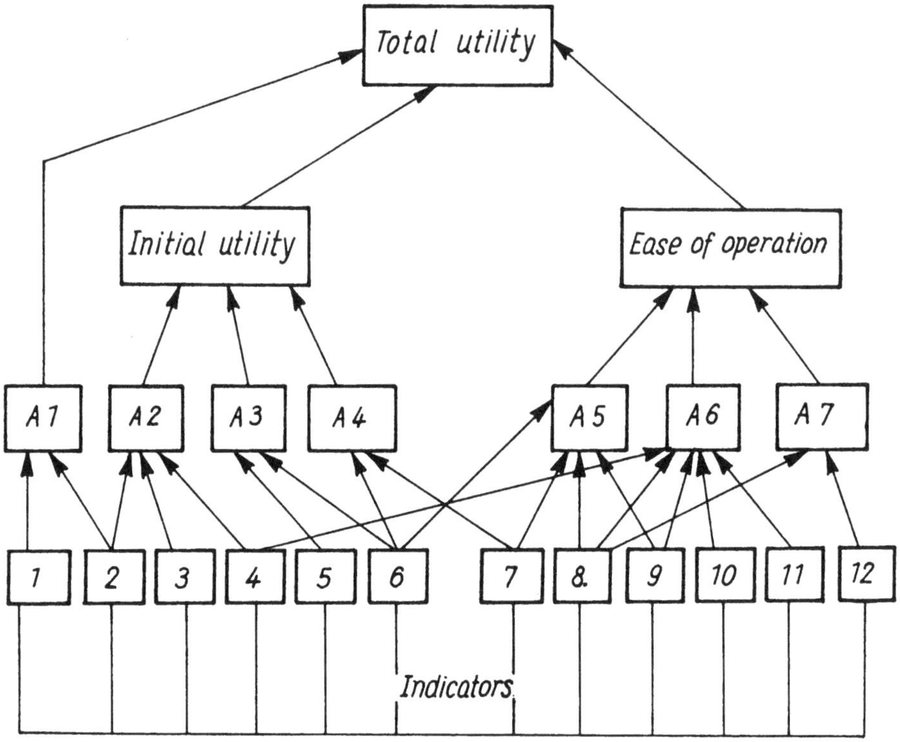

Fig. 2 Hierarchical tree of software properties (Boehm et al.,1978): A1 - mobility; A2 - reliability; A3 - efficiency; A4 - the human factor; A5 - estimability; A6 - comprehensibility; A7 - modifiability; 1 - independence of computer; 2 - completeness; 3 - accuracy; 4 - consistency; 5 - streamlined configuration; 6 - accessibility; 7 - ease of communication; 8 - structured configuration; 9 - high information content; 10 - significance; 11 - openness; 12 - extendibility; 13 - total utility; 14 - initial utility; 15 - ease of operation; 16 - indicators.

6. The simulation experiment design depends on the research objective and the software structure. For economic and technical studies, the structure of Fig. 3 is generally accepted. In marine ecology, however, only very few papers (Vorovich et al., 1981; Patten et al., 1976; Jansson, 1978; Gal'perin, 1983 a-e, 1986 a & b) report software for simulation models of this kind. In this research, the structure of Fig. 3 was used.

7. The simulation is followed by the processing and interpretation of the results.

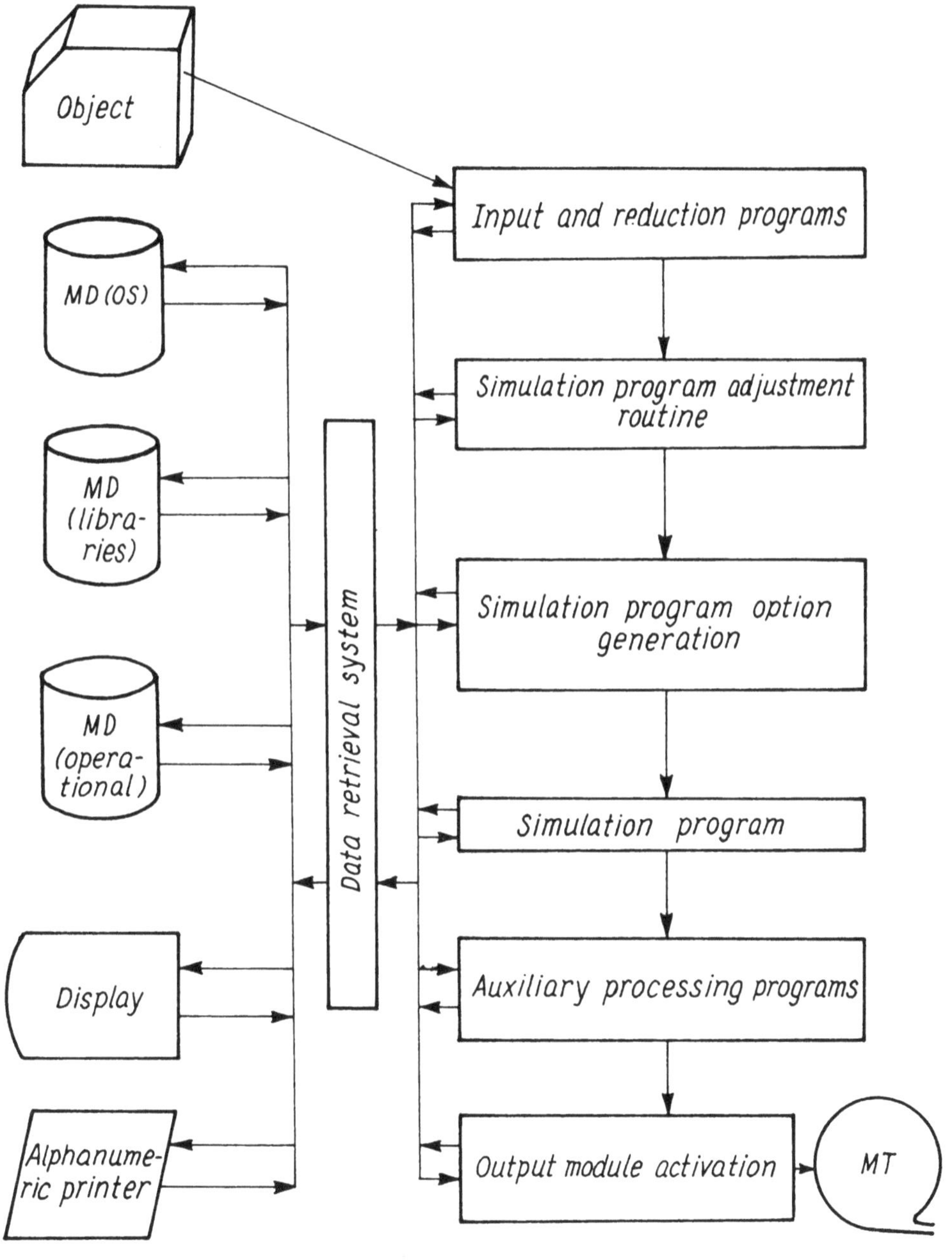

Fig. 3 General structure of software for simulation models (in large computers):
1 - system to be modelled; 2 - input and reduction programs; 3 - MD (OS); 4 - simulation program adjustment programs; 5 - MD (libraries); 6 - data retrieval structure; 7 - simulation program version generation; 8 - MD (operational); 9 - simulation

program; 10 - display; 11 - auxiliary processing programs; 12 - alphanumeric printer; 13 - output module activation; 14 - MT.

8. The results and modules enter the special-purpose data-base of the research.

9. Both, the set of simulation programs and new records in the data-base are documented.

Simulation in Marine Ecology

As noted in the Introduction, the pioneering steps in marine ecology were made by Riley et al. (1949), Kierstead and Slobodkin (1953), Steele and Menzel (1962), and Cushing (1959), who studied plankton communities and their productivity as a function of various abiotic factors. Their models can be referred to as simulating, in a sense, wider than in this paper. The simulation amounted to a computer variation of basic coefficients in the system of equations. The software structure was different from that of Fig. 3 and the principles of simulation introduced above were not maintained. Further extensions are due to Patten in his important papers (1965, 1969), Parker (1968, 1973), Nihoul (1975), and Zeigler (1976). The theoretical aspects of simulating marine ecosystems were discussed by Hannon (1973) and Halfon (1979), while the mathematical formalization of various processes in ecosystems was treated by Ross (1970), Paine (1971), Harris (1972), Pielou (1977), and Platt et al. (1981), to name just a few. In Soviet ecology the most significant results were obtained by Menshutkin (1971), Vinogradov (1973, 1977), Belyaev et al. (1974), Svirezhev and Logofet (1978), and Vorovich et al. (1981). The simulation of production processes in Lake Baikal (Kontorin, 1980), research on the impact made by pollution on the Baltic marine ecosystem (Jansson, 1972; Sjoberg, 1980), and studies of primary production (Cunningham, 1978) were important steps along those lines.

Approaches most closely related with that of this paper were shown by Vorovich et al. (1981), Patten (1976), Zeigler (1984 a and b) and, possibly Levin (1974), Levin and Segel (1982).

Before describing the method of developing a simulation model of the fouling community in the Sea of Azov, the biological concepts and ways to describe the system to be modelled have to be discussed.

1.3. Biological Concepts and Description of the System

Fouling implies a set of animals and vegetation which settle on man-made constructions (Nikitin, 1947; Starostin, 1963; Razumov, 1969; Turpayeva, 1971, 1972) and provide favourable conditions for the development of cruciform bivalve mollusks, lichens, sponges, worms, etc. Besides, moving animals such as crabs, nudibranchs, etc. are also to be found there. According to Turpayeva (1976, 1977) as many as 2,000 species may exist there.

Fouling communities in the southern Soviet seas usually consist of three to five species, two or three of which are attached and one or two unattached, one attached species being dominant as a rule. This feature has also been noted in oligomixt bottom communities (Neyman, 1963; Turpayeva, 1977; Kuznetsov, 1980). In spite of seasonal variations, the structure of fouling communities is dominated by a few species which are usually classified with groups that are systematically and ecologically distinct (Turpayeva, 1967; Kuznetsov, 1980).

Such a community has grown in underwater structures of the metal works on the coast of the Sea of Azov. The basic dominant species are the hydroid, which feeds on the zooplankton; the cruciform balanus, which feeds on the phytoplankton; the colonial infusorian; the predatory nudibranch; and the omnivorous crab. The biocoenosis is dominated by the hydroid, followed by the balanus. The larvae of these species usually settle in late spring and early summer and develop simultaneously. In favourable environment and with a sufficient amount of plankton the hydroid fouls the substrate with a thick cover within a month. Its branching stolons cut the balanus from the food source and this reduces the growth rate of balanus and population. When the infusorian develops on the stolons, the inhibition of the balanus is more severe. In June, the nudibranch joins the community. It settles on the stolons and feeds on the hydroid and infusorian. When it is small, the nudibranch population feeds mainly on the infusorian; when it is large, not only the infusorian is consumed, but also the hydroid is inhibited to such an extent that parts of the colony are deprived of the hydranths (and so of food supplies), wither, and break away. This improves the diet of balanus, which is also enhanced by salinity reduction and the ensuing higher rate of hydroid growth. Shrinkage of the substrate area of the infusorian and nwe nudibranch generations, and the reduced number of live hydranths which the nudibranch feed on, result in a sharper intraspecific competition in the nudibranch population. In July, crab larvae settle in large numbers, juvenile crabs consume the infusoria and then, as they grow, the nudibranches. Large crabs feed on their own juveniles and nudibranch deposits. The crab migrates inside the community and some individuals are carried away with the water current.

These complex processes were studied in actual conditions on a limited area, presuming that, elsewhere, the community develops and functions in a similar way provided that the physical and chemical conditions are the same.

Numerous authors (Beklemishev, 1951; Vinogradov, 1971; Odum, 1971; Margalef, 1968; Caughley,1977; etc.) have noted that the regular nature of seasonal and annual changes in the biota of the community is maintained by the complex pattern of interactions between populations. Usually, however, very few relations determine the processes (Beklemishev, 1951; Shorygin, 1955; Odum, 1971; Patten et al., 1976). Therefore, research in the functioning mechanisms and quantitative estimation of the main interspecific relations by the model would serve the above objectives. This research follows the biological theory of interspecific (synphysiological) relations proposed by Beklemishev (1951).

According to him, all these relations between the populations belong to four types:

1) topic, which act on the environment of the population;

2) trophic, having to do with the processes of nutrition;

3) fabric, arising in the construction (of huts) of the animals;

4) phoric, arising in the transfer of animals of one population to those of other populations.

These relations may be two- and one-way. If it increases the population, the relation is positive, otherwise it is negative. Relations are "direct" if the population of one species makes a direct impact on that of another, and indirect if two populations interact through the impact of one of them on the population of a third species. In a direct relation, the species making an impact is conditioning, and the recipient species is dependent. In the indirect relation there is a modifier which influences the strength of the impact the conditioning species makes on the dependent species (Beklemishev, 1951, 1969; Turpayeva, 1972).

In the fouling community of the Sea of Azov the synphysiological relations influence various aspects of the community development, like the rate of hydroid and balanus growth, the size of nudibranch and crab populations, and the number and size of infusorian colonies. This variety of processes and rates of interspecific relations can be described (Vinberg, 1963; Odum, 1971) in terms of energy flows through the population or variations in the energy value of all the individuals of the species. These indices describe the absolute and relative rate of synphysiological relations and permit a comparison between relations that influence biologically different processes. In addition, other quantities can and, sometimes, must be obtained, such as the relative and absolute values of biological masses and sizes of both the interacting populations themselves and the rates of relations between them (Vinberg, 1962; Naumov, 1963; Khaylov, 1971; Turpayeva, 1972; Odum, 1971).

Energy characteristics of the community cannot be obtained unless the energy and processes with their component populations have been thoroughly studied in compliance with the physical energy conservation law and the ecological law of energy propagation through trophic relating (Vinberg, 1962, 1976; Lyapunov, 1974; Zaika, 1975; Ivlev, 1977; Mesarović, 1968; Odum, 1971; Holme and McIntire, 1971; Paine, 1971; Odum and Odum, 1976).

The biological concept in this research called for initial studies at the organism level. In other words, the energy input-output characteristics must be obtained for one individual or one colony of a population (Turpayeva, 1972, 1976, 1982; Gal'perin, Turpayeva, 1980 a; Gal'perin et al., 1981 a). Following this, the energy flow through the entire population to be modelled is evaluated (Turpayeva, Gal'perin, 1980 b; Gal'perin et al., 1981 c). At this step, the intra-population mechanisms should by no means be neglected. Once this is done, the energy flow through the entire community is evaluated by introducing the impact of interspecific relations (Vinberg, 1976; Lyapunov, 1971; Turpayeva, 1982; Kuenzler, 1961; Hughes, 1970; Dame, 1976; Holme and McIntire, 1971; Odum and Odum, 1976; Patten et al., 1976). What is important is that, despite the requirement that the energy flows in a fouling community and its components have to be explored thoroughly, the energy flows through structural units like the individual, the population, and the entire community, may be represented in a fairly general unified form.

This form, conventional in the bioenergetic ecology (Odum, 1971; Whittaker, 1975), is very useful in integrating structural flowcharts of the simulation model, Fig. 4. As noted above, the energy flow in any closed system must comply with the energy conservation law. In our case, this may be expressed as a system of primitive input-output equations for a structural unit at any level. Using the notation of Fig. 4 we have the following equations

$$IS = P + R + G + F + LE \qquad (1.3.1)$$

$$A = P + R + G + F \qquad (1.3.2)$$

$$AS = P + G \qquad (1.3.3)$$

$$AB = R + F + LE \qquad (1.3.4)$$

Consequently, the compilation of biological concepts, the description of the system to be modelled, and the structural flowchart of the energy of the community components make it possible now to discuss the method of developing a simulation model of the fouling community. Incomplete consumption of certain nutrients in the community may be regarded as energy dissipation. Energy used in moltings of, say, the crab is a case of nonbasic physiological needs.

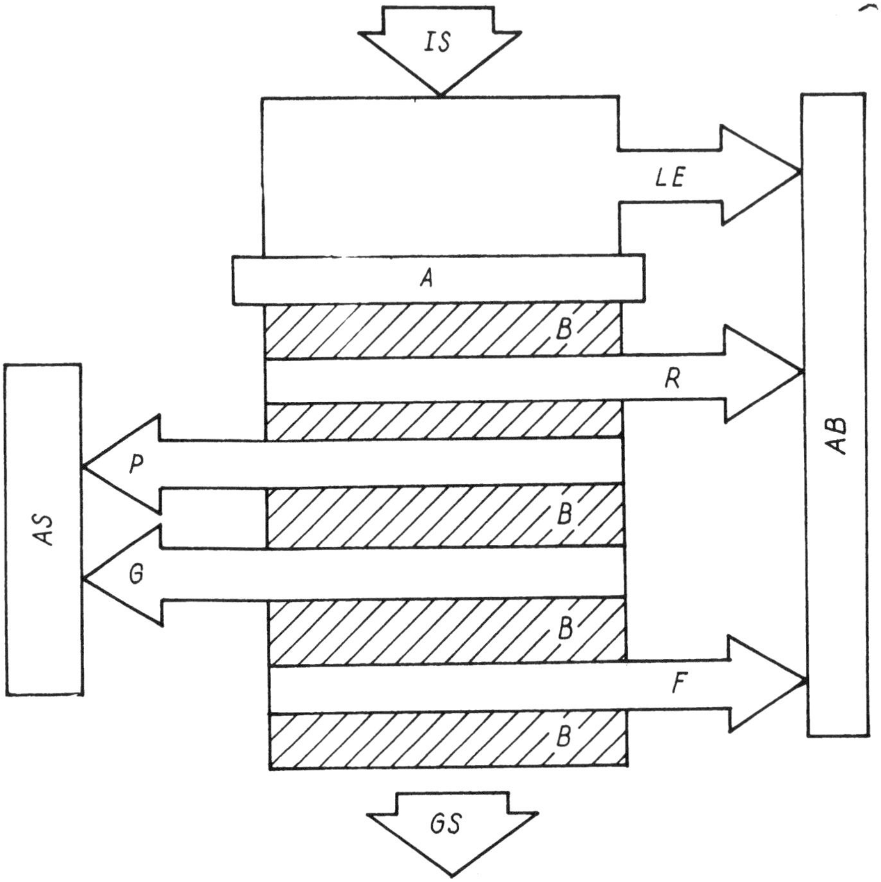

Fig. 4 Unified diagram of energy flow in a structural unit:
IS - solar energy (for producers or preceding trophic level)
LE - unproductive energy losses; R - respiration cost; P - growth cost; G - generation cost; F - cost of other physiological functions; CS - subsequent energy consumer (trophic level); AS - energy accumulation in the system; AB - energy output from the system; B - biomass.

1.4. A Method of Developing a Simulation Model of the Fouling Community

Francis Bacon, the famous philosopher and researcher, insisted on the opinion that getting to know the simplest possible forms of and differences between things was of highest importance because the variety of the world was made up by their diverse combinations. This view fits beautifully into the biological concepts underlying the simulation of the fouling community.

Since the community is investigated bioenergetically, the right-hand sides of the input-output equations (1.3.1) - (1.3.4) are of special interest.

Structuring. The most primitive physiological process expressed in natural (kilograms, number of specimens, etc.) and energy units was chosen to act as a system element. The lowest level in the hierarchy of these processes is associated with the organizational form of life of one specimen; for this reason the system of energy budget is, at the elementary level, the energy budget of one individual. For the set of interspecific relations the elements may be the simplest possible interaction mechanisms and also systemic elements of computing their quantitative characteristics. In compliance with the stages in the design of simulation models, Section 1.2, the system can now be structured (Gal'perin, 1983 a-e).

The system structure is shown in Fig. 5, where the subscripts denote the ordinal number of the model subsystem, which stands for an actual process or community component, while the subscript "j" denotes the type of the interspecific relation (trophic, topic, etc.). The arrows indicate possible relations of some subsystems in the model structure. An element in the energy budget is denoted by e_d; rn_p is the response of a component to an exogenous signal; S_{jc} is a "type j relation" subsystem, and m_y is the mechanism according to which this relation acts. This structure is obviously associated with the unified energy flowchart of community components, Fig. 4, but it also recognizes possible interactions dictated by the energy distributions between components of the actual system and the rate of energy flows.

With the calory acting as an exchange currency in the bioenergetic approach, unified structural flowcharts of the population and interspecific relations are set up, Figures 6 and 7, respectively. These flowcharts are unified for their levels in the hierarchy and no longer hold if steps of an elementary process such as multiplication are investigated. In this case, system structures have to be redeveloped although they remain invariant beginning with a certain level. This is one of the advantages of structural flowcharts. They reduce the efforts in reprogramming the modules when the simulation changes its objective. This structuring preserves the modularity and structure of simulation programs and makes a computer-aided simulation of the system and a generation of simulation model options possible.

Since the above mentioned structural flowcharts are unified in this approach, in describing the simulation models which add up to a simulation set the reader will be referred to appropriate Figures, but for convenience the subsystems in these flowcharts are enumerated. Consequently, software modules will be associated with hierarchy levels.

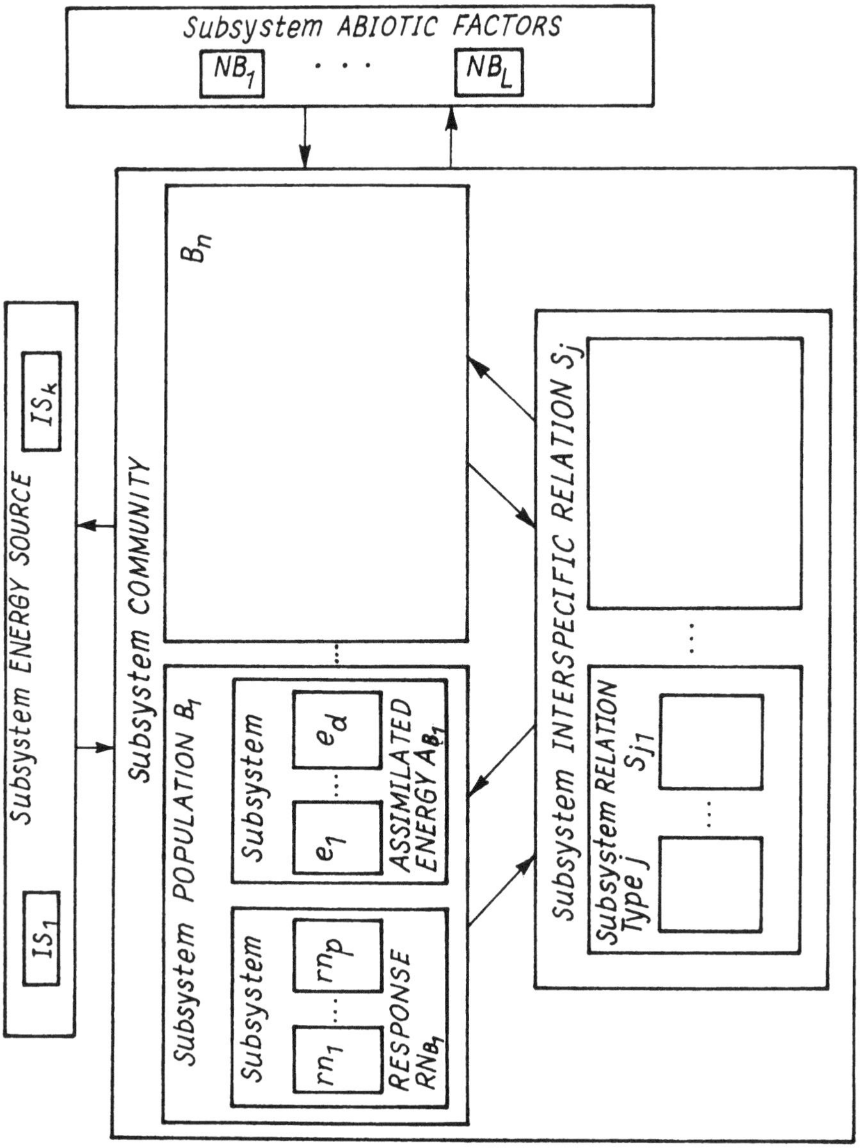

Fig. 5 General structural diagram of the community.

<u>Notes on the Figures 5 - 7</u>

The subsystems and elements in the hierarchy of unified diagrams are enumerated in the following way:

1. First level: elements rn_p, KS , KT_z, m_y, TP_j, T_{sj}, and PR_{sj}.

2. Second level: elements e_d , KR_z, and KF_z and the subsystem Basic relation state.

3. Third level: subsystems Relation type j, Assimilated energy, and Response.

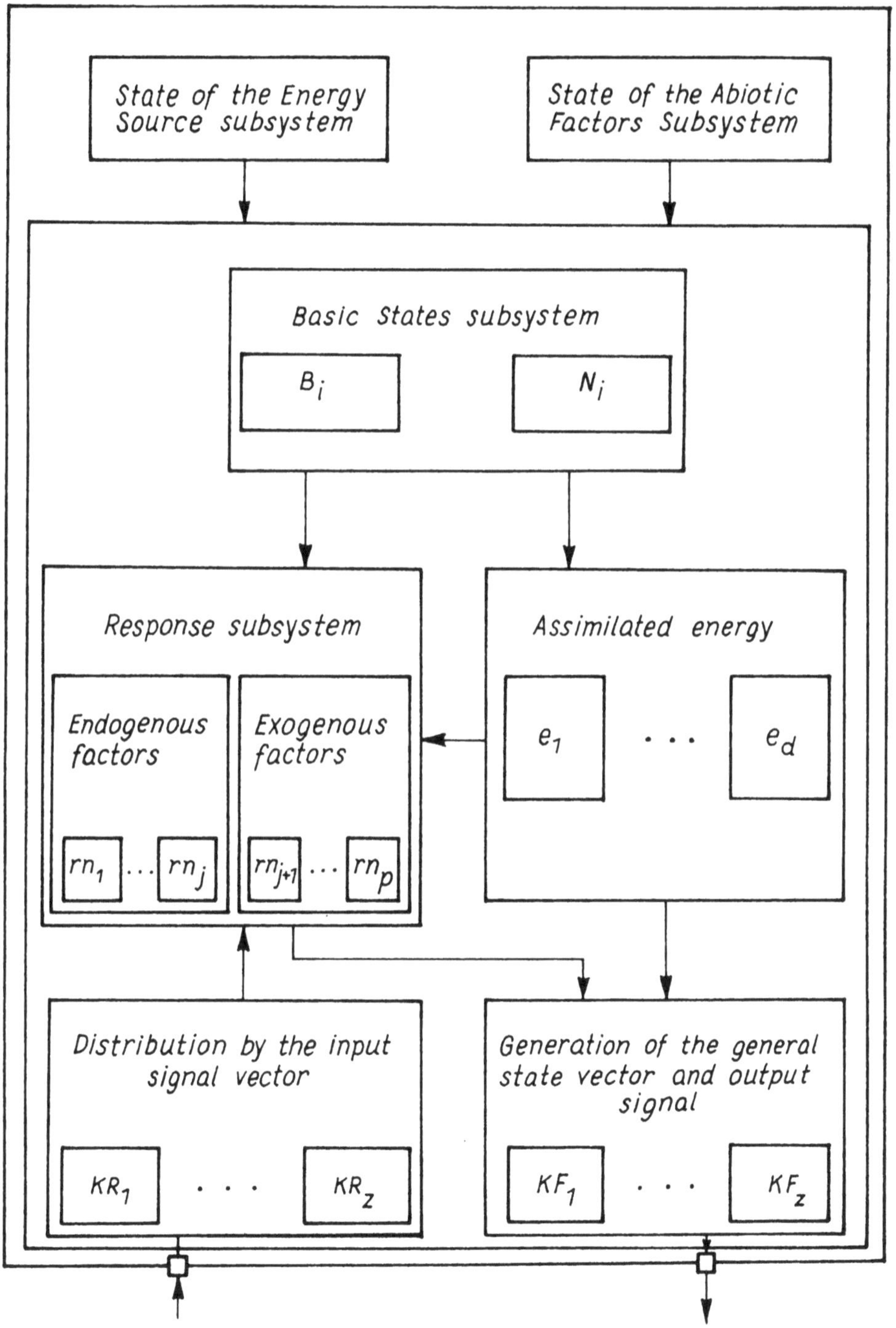

Fig. 6 Unified structural flowchart of POPULATION subsystem (KR_z and KF_z are state vector coordinates, other notation as in Fig. 5).

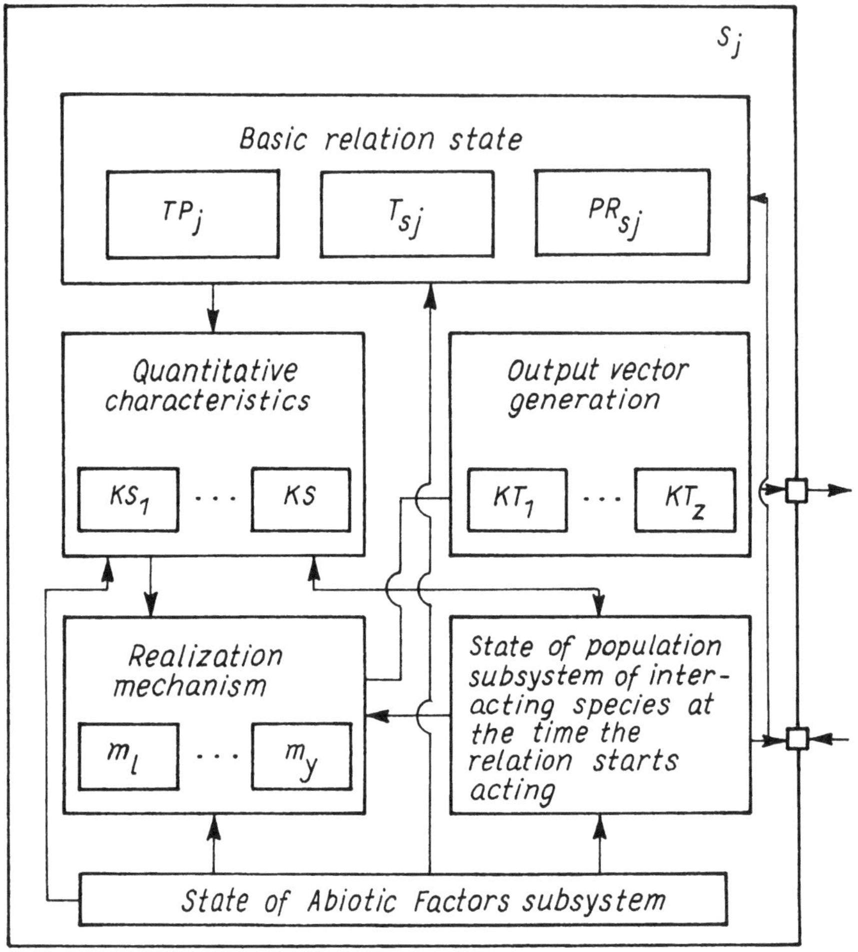

Fig. 7 Generalized unified structural flowchart of INTERSPECIFIC RELATIONS subsystem:
TP_j - element dictating the relation type; T_{sj} - relation action duration; PR_s - relation priority in simulation during step ΔT; KS - quantitative characteristic; KT_z - coordinate of output vector. Other notation as in Fig. 5, SEXTEXT.

4. Fourth level: subsystems Population and Interspecific relation.

5. There is not much sense in having a special level for the Community subsystem because there is just one such system. For this reason, the fifth level covers the subsystems Abiotic factors and Energy source and, conventionally, Control of the model adjustment and of simulation.

Mathematical Formalization

Systematic structuring considering the system to be modelled as a multilevel structure of elements and subsystems of various hierarchical levels dictates the choice of mathematical tools to be used in the simulation model of the fouling community. The mathematical formalization is built around mathematical models representing the functioning of elements in the structured system. In the light of the field experimental data and the fact that most elementary processes included in the interaction of species act over a discrete period, an algorithmic description is important even for subsystems of one hierarchical level. The basic input information of such logical-heuristic algorithms and equations of model elements are empirical relations obtained from the field experiment or reported in the literature. This approach relies on formalized mathematical tools of four levels of generality: (a) descriptions of system elements; (b) more general sequences of process formalization at the population level; (c) still more general sequences of functioning formalization for the entire community; and (d) formalization sequences for the analysis of modelling results. This approach to a mathematical formalization of a system of models to be used in the simulation research of the fouling community may be referred to as the top-down principle. Logical-heuristic algorithms of higher hierarchical levels do not introduce new information into the modelling of this biosystem; in other words, their procedures are formal. Indeed, to obtain the total number of, say, crabs in the community, the numbers of crabs of different sizes have to be obtained and then added up for specific days. Using such an algorithm at the population level, however, maintains the compatibility of unified structural flowcharts and the modularity and structure of programs in the simulation set. The development of mathematical models by the approach of this paper requires a significant amount of properly collected field data (see Introduction and Section 1.3). The conventional approaches amount to deriving a system of equations which use characteristics (coefficients, variables) that are integral in terms of the data on the system (Alekseev, 1973; Alekseev and Polyakova, 1974; Aizatullin, Leonov, 1975; Smith, 1974; Svirezhev and Logofet, 1978). In this approach, special attention has to be given to the resulting systems of equations and, for new objects, a mathematical model has to be developed from scratch (Belyaev, 1978). In studying the interaction of species, various kinds of Volterra equations are extensively used. But the cases in which Volterra models proved adequate were ecologically plain (Svirezhev and Logofet, 1978). On the other hand, Volterra-like systems make it possible to study mathematically the stability of system functioning and the stability over a time of the system structure in the face of various factors (Lyapunov, 1971,

1974; Svirezhev, Logofet, 1978). Sometimes the equations of physical problems are useful. Thus, the movement of insects along a uniform stripe of food was described by the heat conductance equation; when another insect species was introduced, competition started and the model had to be reformulated along the Volterra lines.

The conventional approach is inapplicable to the fouling community because there are no quantitative data on the coefficients or rates of energy flows, or, generally, on interactions of species.

Software

The design of unified structural flowcharts and the down-top mathematical formalization fits into the state-of-art concepts of structural programming in the best way imaginable (Jodan, 1975). The top-down principle is employed in the structural approach to develop software complexes, whereas the debugging of modules starts with the primitve procedures of the lowest hierarchical levels mostly. Special-purpose libraries preserve the modularity of programs. The data to be employed in computation, or the database, also have to be stored in the library on magnetic plates, disks or tapes. Therefore, the software for the simulation of the fouling community has to be structured as in Fig. 3, Section 1.2 in general. The computer implementation of the simulation complex will be described in more detail in Chapter 3. At this point, let us only note that, in this research, an attempt was made to supplement the adjustment of modules and models to a specific variety of simulation by the generation of a certain version of the simulation complex in compliance with a simulation design. The simulation program of a model may be adjusted by specifying the way in which the first hydroids of the colony settle. In one case they were assumed to settle simultaneously and in the model of the entire community over an extensive period (Chapters 2 and 3).

The simulation design envisaged a simulation experiment with both models of individual populations, Chapter 2, and with a model of the entire community. In general terms, designs of simulation experiments included studies of the dynamics of systems as a function of nutrition supply and abiotic factors as well as the impact of the conditions under which the biota develops and functions, depending on the parametrically specified environmental conditions (this is associated with a certain generation of the simulation program).

Assumptions

Like any other method, the one applied in this research for the development of a simulation model proceeds from certain assumptions.

1. To avoid estimating the circadic rhythms and to be able to compute the diurnal energy and other quantities, for the modelling step one day is chosen.

2. With this modelling step, all processes can be assumed to occur without delays. In a general case this assumption is probably not necessarily correct, but it is useful when the discrete step has this length. In the mathematical description and the software implementation of the simulation model the continuous processes can be represented in a form convenient for computer simulation and parallel processes, and their interaction can be reflected. As shown, to make this possible the situation must be modelled within the general modelling step (Gal'perin, 1983 a-e). Sometimes these two objectives can be achieved by using hybrid computing systems and special-purpose modelling languages such as GPSS, SIMULA, SIMSCRIPT, ISL-8, GANIS, ASTRAL, etc. (Kogan, 1978). Primitive ecological-economic systems can also be simulated in this approach by using analogous computers (Lenin and Logvin, 1977).

3. The processes in the biosystem can be quantified and represented consistently in the model in compliance with the biological concept on the possible course of the process and its actual embodiment. Models of interspecific relations are obtained in a consistent way, assuming that the topic relation has to be established as a first priority (which follows from the general biological sense of topic relations) and taking into account that most of the durable relations have a higher priority (Gal'perin, 1983 a-e; Turpayeva, 1972, 1982; Gal'perin and Turpayeva, 1982). Quantitative estimates of potential processes are obtained from models of populations which develop in isolation from each other (such models will be referred to as autonomous in the following). Following this, the mathematical description and software modules with a set of models of interspecific relations lead to estimates of the actual processes in the community. This approach, as noted above, would have been impossible without a field experiment (Turpayeva, 1969, 1972, 1976, 1982), which yielded data on the development of populations in isolation or in a community. The impact of this computation method on the numerical result is discussed in Sections 3.4.

4. The pattern of the autonomous functioning of some element does not necessarily remain the same when it is combined with other elements. Therefore, in the light of the biology of a community, special-purpose additional modules must recognize changes when autonomous functioning is replaced by functioning in combination. Such modules are denoted by Population state recomputation. Among the processes to be updated

in this way are the distribution of the nutrients assimilated by the hydroid for physiological needs (allowance for regeneration) and mechanisms by which the nudibranches and the infusorian populations are restored.

Basic Features of the Method

Briefly, the basic features of the method used for a simulation model of the fouling community are:

1. The functional (blockwise) structure of the object is supplemented with a systematic structure, Figures 5-7.
2. The mathematical formalization is represented in the form of diagrams of four generality levels. A major role is played by the algorithmical form of the description, notably for the population and biocoenotic levels.
3. Special-purpose libraries of modules and data are employed. The software complex is modular and structured. The models of autonomous populations are adjusted to the simulation version, and diverse versions of models are generated, Chapters 2 and 3. The control module of the simulation model generates the specific configuration of the software complex and unifies the input and output.
4. The simulation experiment is performed according to a simulation design specified in advance for biological problems and recognizing the differences both in the community environment and in the nutrition.
5. The behaviour of the biosystem is dynamically forecasted for different conditions and the impact of basic factors is estimated. Such a forecast is made both for autonomous models representing the potential of populations and for populations in the community. The basic characteristics of the entire community are estimated.

Model Description Format

The model and results will be described in the following sequence, designed for clarity and logical consistency:

a) Description of the object.

b) Mathematical formalization, the implementation algorithm.

c) Simulation schedule and quantitative characteristics of the simulation program, description of the functioning of basic models.

d) Adequacy evaluation of the results.

e) Evaluation of the forecast and analysis of the simulation result.

f) Statistical estimation (in conclusion of Chapters 2 and 3).

Chapter 2. SIMULATION STUDY OF THE DEVELOPMENT AND FUNCTIONS OF COMMUNITY SPECIES WITHOUT REGARDING INTERSPECIFIC RELATIONS (AUTONOMOUS MODELS)

2.1. A Hydroid Population Model

The Hydroid

Turpayeva et al. (1977) indicated that coelenterata colonies add up to an intergrated system which controls the shape, size, growth rate, nutrition rates, the distribution of nutrients and the cellular material. The hydroid is a dominant fouling species in the Sea of Azov. The hydroid Perigonimus megas Kinne life cycle can be divided into four stages:
I - development of a juvenile colony from the first polyp to maturity;
II - reproduction of a part of the colony formed in the first stage, its transformation into a somatic colony, and growth of the generative part;
III - reproduction of the generative part, separation of the spawned fraction, and growth of new polyps; IV - gradual reduction and separation of the generative part of the colony. There are usually two periods of intensive growth of hydroid colonies, spring and fall. The hydroid biomass is constant for some time in the summer, followed by a sharp decrease; in the winter it decreases gradually. The periods of growth coincide with the mass development of the copepod, Calanipeda aquae dulcis, which is the primary food of the hydroid. Exponential growth is typical of the hydroid and depends on the food concentration. The first stage takes about 25 days. The physiological peculiarities of hydranths prevent an infinite increase of the ration (r') as the food concentration β (in thousand specimens per cubic meter).

Once the maximal value r'_{max} (β_{max}) is achieved, the ration of hydranth remains constant, and the biomass (W_I) of the colony decreases dramatically (under natural conditions at 1.8 - 2.5 kg/m^2) and by the time the colony is mature for large it does not exceed 3.5 kg/m^2. At the second stage, the colony part formed at the first stage is reproduced, the spawned fraction becomes somatic and the nature of metabolism changes.

Once the second stage is over, the hydroid mass is maintained at approximately the same level, and the growth (W_{III}) at the third stage only compensates for the separation of the spawned part (W_{II}). When the fouling was scraped from water intake installations the Perigonim biomass was invariably found to be nonuniformly distributed and gonophores to be present only in some offsprings. In plankton-rich years the hydroid biomass in the nearest underwater areas varied by 1 - 1.5 kg. This leads to the assumption that only a quarter rather than the whole of the generative part regenerates.

At the fourth stage, the generative part reduces and breaks away because of a shortage of regenerated food for metabolism. This happens because of the dramatic reduction and the subsequent lack of food, Calanipeda. The scale of the break down is proportional to the metabolic energy in energy units.

Mathematical Formalization

The processes that determine the development and life of the biosystem are

1. Exponential growth of polyps at a specific rate which is dependent on the food concentration.
2. Slow-down of the growth of the colony with a maximal ration of hydranths.
3. Four stages in the population growth when the colony is mature and develops as a function of the availability of food.
4. Population regeneration, a part of the population becomes somatic, separation of the regenerated parts as a function of energy metabolism.

As noted above, the hydroid colony grows exponentially at a specific rate, $C_W(\beta)$, which depends on the food concentration β. Turpayeva and Simkina (1975) related the amount of the plankton food with the growth rates of the colony per square meter of the experimental plate. The rate was obtained by the standard technique (Vinberg, 1976), presuming power growth. For the first stage

$$C_W(\beta) = \begin{cases} \alpha_1 \cdot \beta^{\gamma_1} & \beta < \beta_{max}, \\ \text{const} & \beta \geqslant \beta_{max}. \end{cases} \qquad (2.1.1)$$

The parameters of this equation are estimated by the method of least squares and assumed to be $\alpha_1 = 0.1$ and $\gamma_1 = 0.5$. The statistical estimate of the significance of α_1 and γ_1 was obtained from field data, s. 2.5.

To derive the equation for the growth of the colony at the first stage, the slow-down of the population development has to be recognized. Turpayeva et al. (1976) assumed that the biomass the slow-down starts with was 2 kg, the average of several experimental plates. This is true for β being so high that the exponential equation leads to exceeding the maximal biomass of 3.5 kg.

In the equality

$$W_I(t^*) = 2 \text{ kg/m}^2 \qquad (2.1.2)$$

time t^* is that time where the growth slow-down of a colony which grew until that time by an exponential law such as

$$W_I(t) = W_0 \cdot \exp\,(C_w(\beta) \cdot t), \tag{2.1.3}$$

where W_0 is the initial biomass of first polyps, which is, according to (Turpayeva et al., 1976), equal to 0.00001 kg/m^2. After time t , the colony development is described by equation (2.1.3) but the slow-down of the population growth has to be recognized, the population growing during the remainder of the first stage to 3.5 kg/mm^2 on the average, as in

$$W_I(t) = W_0 \cdot \exp\,(C_w(\beta) \cdot t^* + C_w^* \cdot (t - t^*)), \tag{2.1.4}$$

where $C_w^* = \left[\ln\,(W_{max}^*/W_I(t))\right]/(25 - t^*)$ and $W_{max}^* = 3.5$ kg/m^2. According to the field experiment, the energy of diurnal metabolism in one kilogram of a colony of the same age is $T_1 = 65.31 \pm 0.01$ kcal/kg (Turpayeva and Simkina, 1975; Turpayeva et al., 1976). The energy value of one kilogram of the colony was estimated in the experiment as $\Psi_G = 670$ kcal/kg. Now, the energy metabolism effort may be obtained from the equation

$$R_I\,(t) = T_1 \cdot \left[W_I\,(t + 1) - W_I\,(t)\right], \tag{2.1.5}$$

and the effort for increase, from the equation

$$P_I(t) = \Psi_G \cdot \left[W_I\,(t + 1) - W_I\,(t)\right]. \tag{2.1.6}$$

The budget equation for assimilated energy is

$$A_I(t) = P_I\,(t) + R_I\,(t). \tag{2.1.7}$$

The total number of hydranths increases at the first stage by the equation

$$N_I(t) = N_0 \cdot W_I\,(t), \tag{2.1.8}$$

where $N_0 = 865 \cdot 10^4$ is the number of hydranths per kilogram of the hydroid biomass. The individual hydranth ration may be obtained as

$$r' = A_I\,(t)/N_I\,(t). \tag{2.1.9}$$

Vinogradov et al. (1973) and Turpayeva et al. (1976) used the coefficient, which characterizes the extent to which the colony uses the accessible food, as the indicator of feeding regulation. The coefficient was determined as the ratio of the total energy costs of the hydroid obtained from (2.1.7) to its theoretically possible maximal food supply. The latter is obtained as the product of the number of live hydranths and their individual ration (r'). The value of r' has been shown (Turpayeva et al., 1976) to depend only on the number of plankton specimens, and the regulation proceeds by reducing the number of feeding hydranths, or

$$\delta = N_E/N_O, \tag{2.1.10}$$

where

$$N_E = C_w^*/C_w\ (\beta). \tag{2.1.11}$$

In computing the characteristics of the adult hydroid population, the determination of the weight proved to be the most difficult problem because the generative part of the colony, W_{II}, grows in different ways depending on whether the plankton is in good supply. Field studies of the hydroid ecology revealed that the generative part grows and asymptotically approaches the maximally possible biomass, $W_{II\ lim}$, which was found to depend on the food concentration. In planktonrich years, when β exceeds 30, the hydroid biomass on the waterintake installations amounted to 8 kg/cm^2. In plankton-lean years, the hydroid fouling mass was much less (Turpayeva et al., 1977). The constraint on the biomass, the "upper asymptotic value", led to the assumption of an S-shaped pattern of the colony growth at the second and third stages. For this reason the biomass equation was represented as

$$W_{II}(t) = \frac{W_{II\ lim}}{1 + \exp\ (-C_w \cdot t)}. \tag{2.1.12}$$

For the first five modelled days the computation by this equation revealed, however, inconsistency with the field data. To make the biomass curve fit the observation data as well as possible, a control parameter Ω (t) was introduced to the equation,

$$\Omega\ (t) = \begin{cases} \sin\left[(t \cdot \pi)/10\right] & t \leq 5, \\ 1 & t > 5. \end{cases} \tag{2.1.13}$$

Then (2.1.12) rearranges into

$$W_{II}(t) = \frac{W_{II\,lim} \cdot \Omega(t)}{1 + \exp(-C_w \cdot t)}. \qquad (2.1.14)$$

The value of $W_{II\,lim}$ as a function of β was estimated by a conventional statistical technique (Section 2.5) as an exponential function with parameters $\alpha_2 = 0.1$ and $\gamma_2 = 1.1$.

$$W_{II\,lim} = \begin{cases} \alpha_2 \cdot \beta^{\gamma 2} & 20 < \beta \leq 35, \\ \text{const} & \beta > 35. \end{cases} \qquad (2.1.15)$$

Computation by formula (2.1.14) proved satisfactory only with $\beta > 20$. In the $1 \leq \beta \leq 20$ range the energy of food the colony could obtain by feeding all the hydranths is insufficient for a growth of W_{II} by (2.1.14). As there were only a few data on the dynamics of the hydroid with plankton available, we assumed that under these conditions the hydroid, firstly, consumes all the food available, a fact represented by the equality

$$A_{II}(t) = A'_{II}(t), \qquad (2.1.16)$$

and, secondly, the food energy is utilized, first of all, for the respiration and regeneration of the colony, W_I, and only the remainder is spent for a growth W_{II}. In this case the colony grows nearly exponentially, but in computer simulation the rate of growth was found to vary with time. The value of W_{II} is computed as the difference between the total colony biomass W_Σ and its part W_I obtained at the first stage. The total biomass is

$$W_\Sigma(t) = W_I \cdot \exp(\tilde{c}(\beta, t) \cdot t), \qquad (2.1.17)$$

where $\tilde{c}(\beta, t)$ is the growth rate of the colony obtained for specified values of β from equation (2.1.16) because both, A_{II} and A'_{II}, depend on the variable W_Σ, and hence on $\tilde{c}(\beta, t)$.

An analysis of the computation by (2.1.17) and a comparison of its results with the hydroid biomass during plankton-lean periods revealed that this matching is true of the interval $1 \leq \beta \leq 7$ only. Then, the assumption was made that in the interval $7 < \beta \leq 20$ the colony first grows according to (2.1.17) and then according to (2.1.14), and that the growth of the hydroid changes when W_{II} reaches the value at which the ration is sufficient for the colony to respire and grow W_{II} in compliance with (2.1.14).

At the third stage, the computation by (2.1.14) was found to be legitimate in the entire range of β. In this modelled period, about one quarter of the colony formed at the first stage is engaged in regeneration and the regenerated fractions break away. Computing by (2.1.14), the growth of new parts is presumed to proceed sequentially as the spawned areas separate, or

$$W_{III\ lim} \approx 0.25\ W_{II}. \tag{2.1.18}$$

The total hydroid weight is obtained from the equation

$$W_{\Sigma} = W_I + W_{II} + W_{III}. \tag{2.1.19}$$

The population increase and the required energy costs are computed by (2.1.6). The rate of energy metabolism of some stolons with one hydranth is given by the equation

$$\theta = T_1 : N_0 \approx 7.55 \cdot 10^{-6}. \tag{2.1.20}$$

This quantity is essential in computing the total respiration costs of the colony. Hydroid colonies multiply by planulae, which form in female gonophores. Every gonophore takes four days to develop, which is followed by spawning. The energy value of a gonophore varies with time (Turpayeva and Simkina, 1975) and is approximately represented by the equation

$$\xi(t) = \begin{cases} \alpha_3 \cdot t_\xi^{\gamma_3} & t_\xi \le 4, \\ 0 & t_\xi > 4, \end{cases} \tag{2.1.21}$$

where the parameters $\alpha_3 = 0.5 \cdot 10^{-3}$ and $\gamma_3 = 1.28$ are obtained by the conventional method of least squares (concerning the significance see Section 2.5). The number of gonophores in areas taking up 100 hydranths increases by the equation

$$D(t) = \begin{cases} \alpha_4 \cdot \exp(\gamma_4 \cdot t) & D(t) \le \widetilde{D}, \\ \widetilde{D} & D(t) > \widetilde{D}, \end{cases} \tag{2.1.22}$$

where $\widetilde{D}$ is the maximal number of gonophores in an area and equal to 1.000 because an average of 10 gonophores develop under one hydranth. The coefficients α_4 and γ_4 were obtained experimentally (Simkina, Turpayeva, 1980) for various temperatures: at 20 °C: $\alpha_4 = 0.26$ and $\gamma_4 = 0.85$; at 24 °C: $\alpha_4 = 0.47$ and $\gamma_4 = 1.2$, Section 2.5. These data lead to the duration t_k of the spawning period. The regeneration energy is considered as the sum of spawning energy costs of gonophores which are at different stages

$$G(t) = \Delta W_H \cdot N_0 \cdot \sum_{j=1}^{t_k^{II}} \sum_{i=1}^{t_k^{III}} \Delta \xi_i^{(j)} (t) \cdot \Delta D_j^{(i)} (t), \qquad (2.1.23)$$

where $G(t)$ is the food energy spent for the regeneration of a part of the colony weighing ΔW_H kg; $\Delta\xi_i^{(j)}(t)$ is the coloricity of gonophores in the j-th group on the i-th day of their development; $\Delta D_j^{(i)}(t)$ is the number of gonophores in the j-th group on the specified day of their development; t_k^{II} and t_k^{III} are equal to 10 and 4 at the second stage (20 °C), and to 7 and 4 at the third stage (24 °C). The number of spawned gonophores $D^*(t)$ was determined as

$$D^*(t) = \begin{cases} 0 & t < 4, \\ \alpha_4 \cdot \exp(\gamma_4 \cdot (t - 4)) & 4 \leq t = t_k + 4, \\ \tilde{D} & D^*(t) > \tilde{D}. \end{cases} \qquad (2.1.24)$$

Following the spawning, the hydranths in the neighbourhood of which the gonophores developed reduce. The number of reduced hydranths is given by the equation

$$n(t) = \Delta W_H \cdot N_0 \cdot D^*(t), \qquad (2.1.25)$$

and the total number of live hydranths in a colony is

$$N(t) = W_\Sigma (t) \cdot N_0 - n(t). \qquad (2.1.26)$$

The spawning and reduction entail a variation in the metabolic energy rate. After the reduction, the rate of oxygen consumption by the spawned offspring is one tenth of what it used to be. The metabolic energy cost in one reduced hydranth is $\theta \cdot 10^{-1}$ and the total energy metabolism cost may be obtained from the equation

$$R(t) = \theta \cdot \left[N(t) + n(t) \cdot 10^{-1} \right]. \qquad (2.1.27)$$

The assimilated food was computed as the sum of growth, respiration, and regeneration costs

$$A(t) = P(t) + R(t) + G(t), \qquad (2.1.28)$$

conventionally referred to as the real ration, and as the product of the number of hydranths by the individual ration (r'), conventionally referred to as the potential ration (A') of the colony

$$A'(t) = N(t) \cdot r'. \quad (2.1.29)$$

Then δ can be obtained as

$$\delta = A(t)/A'(t). \quad (2.1.30)$$

This value shows what fraction of the available energy is utilized by the colony.

What is important is that in simulating the third stage the relations and equations for the growth, metabolism, and regeneration have a new model time reference for the substages to allow for the assumption on successive regeneration and growth of new parts of the colony after the separation of spawned parts. In one substage of the third stage, a weight equal to about one quarter of the biomass grown at the second stage was removed.

The Simulation Program

The simulation program consisted of seven basic and two auxiliary modules. The first one included: option adjustment (M1), specification of initial simulation conditions and computation of general population characteristics (M2), four simulation models, one for every stage (M3 - M6), and population regeneration (M7). The latter were: interaction (input/output) with external memories (M8), and print-out and plotting (M9). The total running time of the simulation program was 5.61, the net CPU-time 1.46, and the operation without print-out took 3.24 minutes. The computer memory needs are shown in Table 1.

The editing time was reduced and the quantity transfer between the eight modules simplified by using two common locations. The one stored the values of the current model day T and the values of the auxiliary parameter, and the other the values of equation constants and variables. The model time of one run was 100 days.

Simulation Design

Two series of 16 runs were used to study and quantitatively estimate the energy flows through the hydroid population.

The main simulation series were supplemented with two additional series. The main series were different in the settlement period and in the sequence of growing hydroid polyp branches. In one case the branches were assumed to grow from first polyps practically simultaneously. In

Table 1 Computer memory allocation for the simulation program in a hydroid population model

Characteristic Customer	Number of FORTRAN statements	Core memory, kbytes
M1	43	56
M2	126	12
M3	59	3
M4	77	4
M5	65	3
M6	20	1
M7	40	2
M8	56	3
M9	120	6
Editing	-	108
Execution	-	104
The entire program (initial module)	511	90

another, a protracted (10 days) settlement process and, hence, a certain sequence of branch growth were assumed. The two additional series were different in the modelled salinity S in permille, 5 $^{0}/00$ and 8 $^{0}/00$, respectively. The Calanipeda concentration per cubic meter of water, (β), ranged between 1 and 300 in the model, the step, $\Delta\beta$, being different for different runs; with $1 \leq \beta \leq 10$ it was $\Delta\beta = 1$; with $10 < \beta \leq 100$ it was $\Delta\beta = 10$; with $100 < \beta \leq 300$ it was $\Delta\beta = 50$.

The Functioning of the Main Modules

Morphologically, the simulation program embodied the structural diagrams of the object, Section 1.4. It did not seem desirable, however, to have a separate module for every elementary process in the biosystem in order to avoid overdrafting the core memory and the program run time. In particular, the growth of hydroid population can be obtained from one equation, (2.1.6), whereas having a separate module for this computation would require calling the module, the allocation of some of external memory of its storage, etc. Therefore Table 2, in which the functioning of the main modules is described, indicates the numbers of equations used in the module, the hierarchical level of the mathematical tool in the hierarchy of this research (Section 1.4), and the hierarchical level in the associated structural diagram, Section 1.4. The main operations to be performed by modules that are operated by logical-heuristic algorithms are listed.

Table 2

Module	Main equations or operations	Level in hierarchiy of mathematical tools	Level in hierarchy of structural flowchart
M1	Adjustment, generation of version attribute, i.e. extended settlement or otherwise, generation of salinity attribute and modelling period, call of M2, print-out.	IV	As M1 is control module, conditionally V
M2	Access of library data, writing it in common memory locations, generation and allocation of common locations, arrangement of cycle of model days, call of simulation modules, computation of general population characteristics by eqns. (2.1.7), (2.1.19), and (2.1.28) for different simulation options and relative development times.	II	IV
M3	Simulation of population development at the first period of hydroid life by eqns. (2.1.3) - (2.1.6), (2.1.10), and (2.1.11) then recomputation of the internal time of the first stage from the total model time; print-out.	I, II	I - III
M4	Simulation of the second stage in hydroid life by eqns. (2.1.12) - (2.1.27) and then recomputation of the internal time of the second stage from the total model time; print-out.	I, II	I - II
M5	The model algorithm is analogous M4, but allows for various time and temperature conditions. The same equations with other variables; print-out.	I, II	I - III
M6	Simulation of hydroid functioning without food, or separation of colony parts to offset metabolic energy costs (2.1.20), (2.1.26) and (2.1.27).	I, II	I - IV
M7	Simulation of population regeneration by eqns. (2.1.21) - (2.1.25), then transmission through common locations of specific parameters as a function of the relative stage time and abiotic conditions; print-out.	I	I, II

Adequacy of the Model and Basic Numerical Results

The adequacy of modelling results was tested in three modelled situations with $\beta = 1$, $\beta = 10$, and $\beta = 40$ (Figures 8, 9, and 10, respectively). In autonomous models the case of salinity of 8 $^0/00$ was represented. A lower salinity was used in simulating the development of a population in a community. In an autonomous model, however, appropriate modelling was also carried out. Quantitative data (the hydroid biomass) from a field experiment were available for the case of simultaneous settling of first polyps. Data on their extended settling were also obtained. This is why the adequacy of quantitative results is shown in Fig. 10 for simultaneous settling. Later in the research the extended settlement was assumed not to change the essential nature of the growth and functioning of an autonomous population. Since such a settling period was observed mainly in case of the hydroid in a community, the numerical results obtained from the simulation of this version will be reported in Chapter 3.

In the Figures 8-10 the time is shown for the second and third stages although the model represents the "internal" time of population development. All the results are shown for one square meter of the plate.

At the fourth stage, parts of the colony break away as a function of energy metabolism costs. This process is not especially significant and so the biomass curve will be shown in Chapter 3, when this process affects other species.

With $\beta = 1$, the total biomass W_{Σ} at the second stage increases to 0.00048 kg/m^2 in compliance with (2.1.17), and at the third stage to 0.00055 kg/m^2 in compliance with (2.1.14). This is nearly what was recorded in 1972 (Partaly, 1974), when plankton was in very short supply. At the very end of the third stage, the spawned parts of W_{II} break away and W_{Σ} falls to 0.00048 kg/m^2, Fig. 8. The number of hydranths N increases at the second stage with W_{II} to 3.500 and after the regeneration of W_I decreases to 3.200. At the third stage, N_{Σ} increases to 3.800 and falls to 3.300 after the regeneration of one quarter of W_{II}. The real ration of the colony, A(t), Fig. 8b, at the second stage increases from 0.018 to 0.057 kcal/day and then falls to vary near 0.035 kcal/day at the third stage, and has two peaks, one of which (15th to 18th and 26th to 29th day) is caused by increasing W_{III} and the other (21st to 23rd and 32nd to 34th day) by regeneration of a part of W_{II}. At the second stage, $\delta = 1$ because the hydroid consumes whatever food is available. At the third stage, the nutrition is controlled and δ ranges between 0.5 and 0.7, increasing as the colony grows and regeneration occurs.

The fractions of energy in the real rations that are spent for growth, respiration, and regeneration, Fig. 8c, vary with time. In the first half of the second stage, 53 per cent of the energy is consumed by the growth and 47 per cent by metabolism; at the second stage a maximum of 16 per cent is spent for the regeneration of W_I and the percentage for the growth decreases. At the third stage as much as 60 to 80 per cent is consumed by metabolism, in particular respiration of the reduced part of the colony (for W_I, one per cent); in the beginning of this

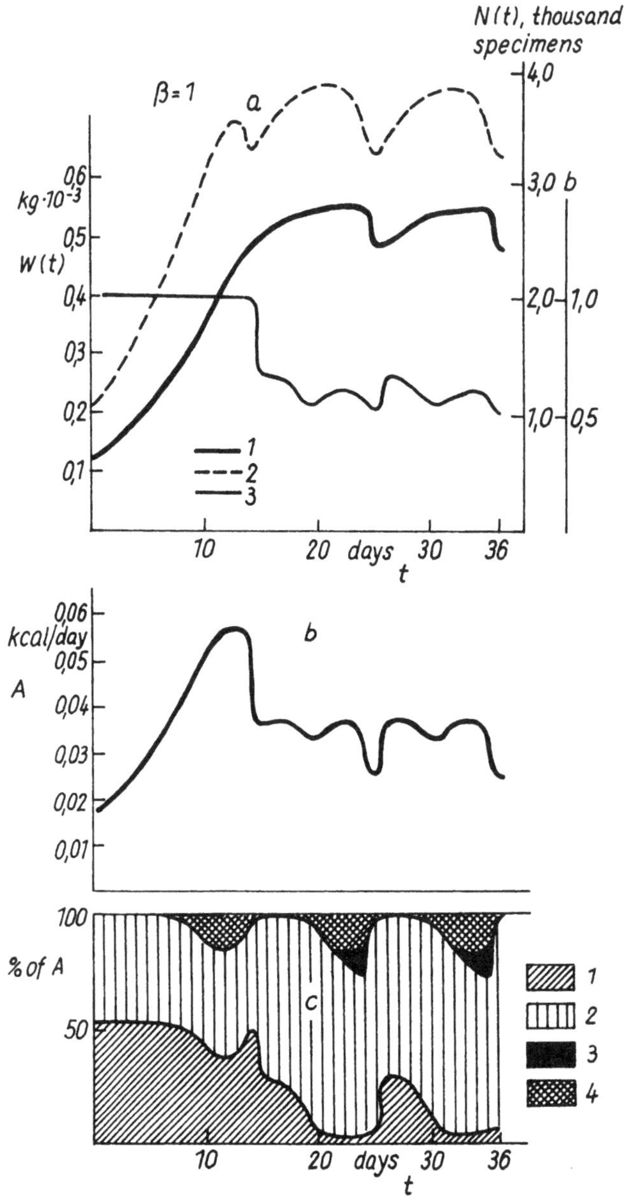

Fig. 8 Characteristics of model perigonimus polulation versus time with β = 1:
a) 1 - total weight W_{Σ} (t); 2 - total number of live hydranths N(t); 3 - function δ (t);
b) Population ration A(t) obtained from input - output equation (2.1.28);
c) Relative amount of energy spent in energy processes: 1 - growth; 2 - energy metabolism of population part with live hydranths; 3 - the same, with reduced hydranths;

4 - regeneration; t - model time of the second and third stages represented for convenience as a continuous scale starting with t = 1.

stage, growth consumes up to 28 per cent of the total energy, and regeneration up to 18 per cent at the end of the stage.

With β = 10 at the second stage, the hydroid develops during the first eight days according to (2.1.17) and then according to (2.1.14); at the second stage, W_{Σ} increases to 1.22 and at the third stage, to 1.51 kg. These values of W_{Σ} are almost equal to the biomass of the Perigonimus (15 g/dm^2) recorded in 1921, with the number of Calanipeda about 10.000 specimens/m^3 (Partaly, 1974). At the second stage N(t) runs into 10,500,000 specimens and at the third stage it first increases to 12,700,000 specimens and, after the spawning, decreases to 10,400,000 specimens, Fig. 9a. At the second stage A(t) increases nearly exponentially (Fig. 9b), to 168 kcal/day during the first eight days, and then decreases to 90 - 83 kcal/day; at the third stage the values of A(t) are higher than at the second stage. During the first eight days of the second stage, growth consumes up to 80 per cent of the energy as assimilated food, Fig. 9a, whereas early in the third stage, only 30 per cent. The energy consumed by regeneration at the second stage is barely above one per cent and at the third stage increases to 16 per cent.

With β = 40, the hydroid population develops in compliance with (2.1.14) from 3.5 to 7.5 kg at the second stage and to 8 kg at the third stage, Fig. 10a. The number of hydranths increases during the first half of the second stage from 30,000,000 to 64,000,000; after the spawning, W_I decreases to 35,000,000 species. At the third stage, N_{Σ} increases to 43,000,000. The value of A(t) during the intensive growth of W_{II} is as high as 1,000 kcal and then falls to nearly 500, while on the 10th - 13th day of the model time (regeneration of W_I) it is as high as 700 kcal/day. At the third stage, A(t) is smaller than at the second stage and ranges between 250 and 500 kcal, increasing during the growth and regeneration periods, Fig. 10b. The value of δ equals 0.32 in the early second stage and then decreases to vary in the 0.12 - 0.17 range, increasing during the growth and regeneration. With β = 40 during nearly the entire model time, 70 to 90 per cent of the energy is consumed by respiration, Fig. 10c, in particular five per cent by the respiration of W_I and parts of W_{II} with reduced hydranths. With the intensive increase of the biomass, growth consumes 66 per cent of the food energy at the second stage and 35 per cent at the third stage. At the second stage regeneration consumes about 33 per cent of the energy A(t) and at the third stage, half that amount.

The total energy flow through modelled populations during the second and third stages, and the values of elements in the hydroid energy balance vary with the food supply. As it increases from 1 to 40, the bio-

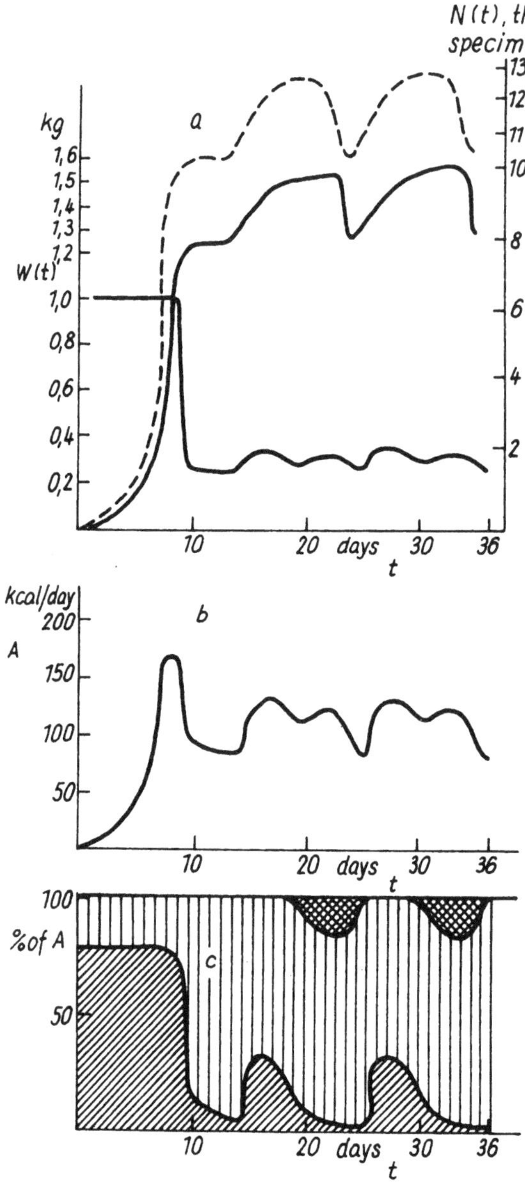

Fig. 9 Model population characteristics versus time with β = 10. The notation is as in Fig. 8.

mass of the modelled population at the second stage increases almost 14,000-fold, and the energy flow 16,000-fold; at the third stage the biomass increases 15,000-fold and the energy flow 11,500-fold. A further increase of β from 40 to 300 does not change the biomass of the modelled population, while the energy flow increases at both stages by as little as one per cent. Most of the food energy is spent for the growth (80 per cent of A), and the smallest percentage for respiration and regen-

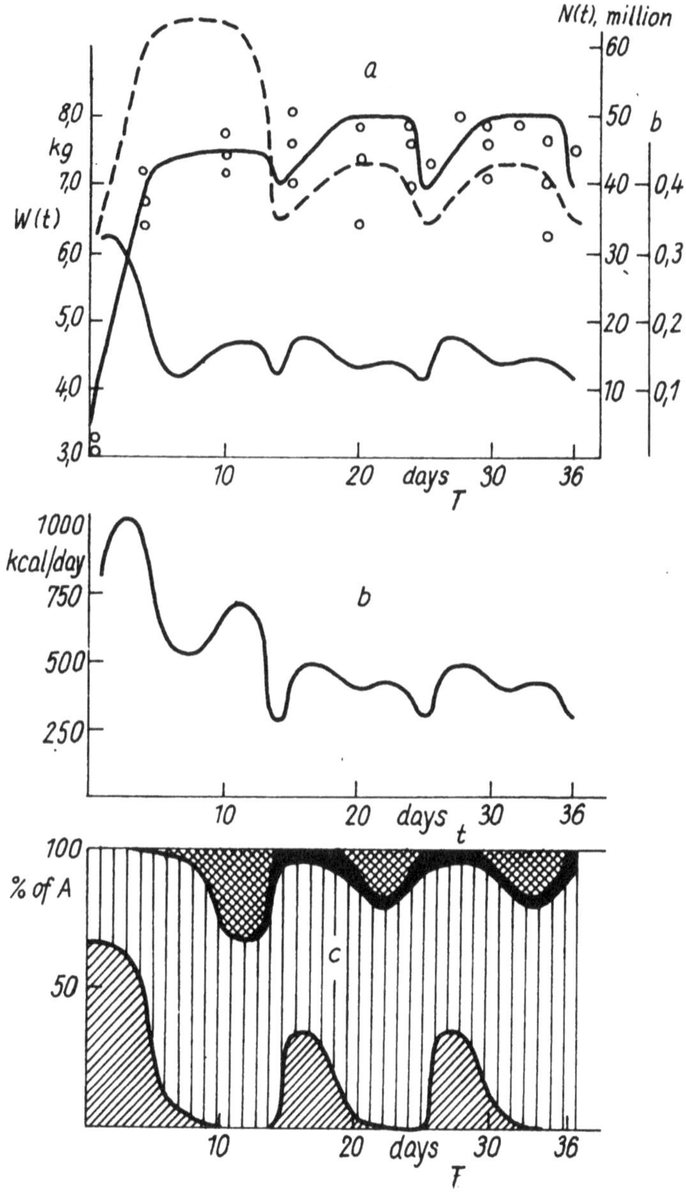

Fig. 10 Model population characteristics versus time with $\beta = 40$. The notation is as in Fig. 8. Circles denote the biomass of hydroid fouling in water-intake facilities, in 1961 (Turpayeva, Gal'perin, Simkina, 1977).

eration with $\beta = 7$. The increase and decrease of the number of plankton specimens entails an increase of the fraction of energy for respiration and regeneration, and the fraction of energy consumed by the growth is reduced. At the third stage the ratio of these fractions in the entire range of β is almost invariable, the growth consuming less and the respiration more energy than at the second stage.

As the supply of plankton varies, so does the utilization of the available food energy by the hydroid. In the $1 \leq \beta \leq 7$ range, the nutrition at the second stage is not controlled and the hydroid utilizes whatever food is available ($\delta = 1$). In the $8 \leq \beta \leq 20$ range, control starts at the second stage, the lower β the earlier, while the value of δ averages between 1 and 0.3, or 100 to 30 per cent of the available energy is utilized. In the $30 \leq \beta \leq 300$ range, control starts at the first stage, while at the second stage the colony utilizes 22 to 4.4 per cent of the available food. At the third stage the food supply is controlled by the utilization (on the average 60 to four per cent) of the available energy.

Now, certain conclusions can be formulated. Firstly, the model showed good fit with the quantitative data not used in its development, Fig. 10. Secondly, for simulation versions for which specific data were not available, the simulation experiment data and that of other papers on the hydroid (e.g. Partaly, 1974) were compared and in no option the computed data differed from the reported ones for similar conditions by more than 15 to 25 per cent. Thirdly, the pattern of the computed curves of hydroid biomass fitted the biological ideas on the qualitative pattern of the colony growth (Turpayeva and Simkina, 1975; Turpayeva et al., 1976, 1977).

These conclusions suggest that the model is adequate and highly sensitive to variations of the rate of growth, a very important parameter. The model also allows for abiotic factors. The modular software corresponding to the structural flowchart made it possible to integrate a simulation program into the model of a fouling community.

2.2. A Balanus Population Model

The balanus improvisus Darwin is an abundant form of fouling found on hydraulic structures in the north-western part of the Bay of Taganrog. The balanus population in that area develops according to a seasonal pattern. The larvae usually settle in the spring (April and May) and in the fall (August and September). The lifetime of the spring generation is 30 to 40 days, during which it grows and spawns. The settling rate of larvae varies from year to year (Turpayeva and Gal'perin, 1980a). In case of a low settling rate, uncrowded populations emerge where the individuals do not interfere with one another's growth. If the rate is high, they inhibit each other in crowded populations. When the abiotic environment changes, multi-layer balanus populations emerge, where the intraspecific competition is very strong and the substrate area is completely covered by the barnacles.

The balanus population model reflects the results of field observations on the number and size variations for various kinds of settlements on experimental plates (Turpayeva and Simkina, 1975; Turpayeva and Gal'perin, 1980 a & b).

The following basic results on the population activities were mathematically formalized:

1. In uncrowded as well as crowded single-layer settlements the barnacles are of the same age and are uniformly distributed over the plate surface.

2. In an uncrowded settlement the barnacles do not contact or interfere with each other until the last day of the model period. There are 226 specimens/dm^2 in such settlements.

3. In a crowded settlement the substrate surface is completely covered and, as they grow, some specimens die.

4. In response to changed abiotic factors the barnacles may settle in numerous layers. In this case, the unit for calculation is the number of larvae settling on the experimental plate in ten days of observation. The growth rate may be determined indirectly by observing experimentally the variations of the population weight.

5. If the multi-layer population reaches a critical value NC, intraspecific competition ensues, some specimens die and the remaining ones grow at a lower rate.

6. When the balanus fouls the experimental plates, where both balanus and the hydroid exist, the hydroid population is inhibited; the hydroid colony growth rate decreases and the time at which the interaction of species starts leading to that decrease is dictated by the growth rate of the population that makes a multi-layer settlement.

7. In uncrowded as well as crowded settlements the barnacles are found (Turpayeva and Simkina, 1975; et al., 1977) to grow exponentially. For a multi-layer settlement experimental data are available on the average weight of a specimen on the plate.

Mathematical Formalization

The ecologists discovered that in both uncrowded and crowded settlements the growth of a specimen may be approximately expressed as an exponential function of the food concentration, α being the number of phytoplankton cells in thousands per cubic meter,

$$\omega_B(t) = \omega_0 \cdot \exp(CW_B(\alpha) \cdot t), \qquad (2.2.1)$$

where ω_0 before the maturity time t^* denotes the initial weight of the specimen, which is $\omega_B(t^*)$ after that time; CW_B is the specimen growth rate which was experimentally found to change after t^*, Table 2. The parameters of (2.2.1) were computed by the standard method of least squares, Section 2.5.

Table 3 Values of the parameters in growth equation (2.2.1) for uncrowded and crowded settlements (for control $\alpha \approx 180$)

Parameter	ω_0		CW_B		t^*
Settlement	$t \leq t^*$	$t > t^*$	$t \leq t^*$	$t > t^*$	
Uncrowded	0.0092	3.65	0.60	0.08	11
Crowded (single-layer)	0.0092	8.5	0.51	0.01	14

As noted in the description of the system to be modelled, for a multilayer settlement only data on variations of the average weight of a specimen on an experimental plate were obtained as a function of the population age.

This function is

$$\omega_B(t) = a_1 \cdot t^{b_1}. \qquad (2.2.2)$$

Four pairs of values of a_1 and b_1 were statistically identified as a function of the settlement of larvae and the conditions in which the settlement emerged (number of ten-day periods of observation), Table 4.

Now the energy costs of the specimen growth are

$$P_B(t) = \Psi_B \cdot \left[\omega_B (t + 1) - \omega_B (t)\right], \qquad (2.2.3)$$

where $\Psi_B = 0.71$ kcal/mg is the energy value per unit weight of a growing barnacle. The energy costs PEX_B of moltings were found in field experimentation (Turpayeva, and Simkina, 1975; Turpayeva, and Gal'perin, 1980 a & b) to be nearly constant, five per cent of the energy value of a specimen. The metabolism cost exponentially grows with the specimen's weight

$$R_B = \alpha_5 \cdot \omega_B^{\gamma_5}, \qquad (2.2.4)$$

where $\alpha_5 = 0.1$ and $\gamma_5 = 0.8$ with significance by the Student criterion at the 2.5 per cent level. The confidence interval for α_5 is $\pm$ 0.004, and $\pm$ 0.04 for γ_5.

The budget equation for a specimen is

$$A_B = P_B + PEX_B + R_B + G_B. \qquad (2.2.5)$$

Note that the regeneration energy can be deduced from (2.2.5) if the value of A_B is known. Once the value of the daily food energy assimilated by the specimen during intensive balanus growth (1st to 11th day in an uncrowded settlement and 1st to 14th day in a crowded settlement) is determined, the food variation can be obtained as a function of the balanus weight. For both uncrowded and crowded settlements this dependence was found to be a power function with significance of parameters at the five per cent level by the Student criterion

$$A_B = \alpha_6 \cdot \omega_B^{\gamma_6}, \qquad (2.2.6)$$

where $\alpha_6 = 0.72$ and $\gamma_6 = 0.96$ for an uncrowded settlement, and $\alpha_6 = 0.58$ and $\gamma_6 = 0.95$ for a crowded settlement.

Since the dependence (2.2.6) must, for biological reasons, also hold for multiplying animals, the energy costs of multiplication (G_B kcal/day) may be derived from the equation (Turpayeva and Simkina, 1975; Turpayeva and Gal'perin, 1980 a)

$$G_B = A_B - (P_B + R_B + PEX_B). \qquad (2.2.7)$$

In order to obtain numerical values of the population energy balance from the equations (2.2.2) - (2.2.7), the number of specimens in the settlement type should be allowed for variation. In an uncrowded settlement the population is invariable and this case is not a challenge. In a crowded settlement the population variation on the experimentally submerged plates is described as a function of the specimen weight

$$N_B = \begin{cases} N_0^{(1)} \cdot \omega_B^{-\gamma_7} & \omega_0 \le \omega_B \le \omega_{max}, \\ N_0^{(2)} \cdot \omega_B^{-\gamma_8} & \omega_{max} \le \omega_B \le \tilde{\omega}_g, \\ \text{const} & \omega_B > \tilde{\omega}_g, \end{cases} \qquad (2.2.8)$$

where $N_0^{(1)}$ and $N_0^{(2)}$ are initial numbers of specimens on the plates (Turpayeva and Gal'perin, 1980 b); the parameters $\gamma_7 = 0.4$ and $\gamma_8 = 0.6$ represent the natural elimination of specimens; $\tilde{\omega}_g$ is the weight of adult barnacles (for statistical estimates see Section 2.5); and ω_{max} is the maximal weight of a specimen with which the huts are contiguous $\omega_g = 9.5$ and $\omega_{max} = 0.5$).

For a multi-layer settlement the number of barnacles in a population may be obtained from the data on larvae settlement, Table 4, and specimen growth equation (2.2.2). The equation of population is

$$N_B = \int_\tau \left[NS_\tau(T) - NG_\tau(t_\tau) \right] d\tau, \qquad (2.2.9)$$

Table 4 Parameters of equations modelling a balanus population in a multi-layer settlement

Quality Option	Number of larvae settled in 10 days; $(\overline{NS} \cdot 10^{-3})$						a_1	a_2	b_1
I	400	200	300	-	-	-	0.0017	7.85	2.46
II	6	9	18	30	50	-	0.0017	1.72	2.51
III	50	180	150	150	150	35	0.0015	2.37	2.40
IV	100	100	150	100	200	150	0.0016	4.39	2.35
Run	1	2	3	4	5	6			

where NS_τ is the number of settled and developing larvae; NG_τ is the number of barnacles dying of natural causes: t_τ is the life span of the τ-th age bracket of the balanus and T is the model time. In discrete time this equation is

$$N_B = \sum_{\tau=1}^{T} \left[NS_\tau(T) - NG_\tau(t_\tau) \right], \tag{2.2.10}$$

where T is the total model time. With an allowance for intraspecies competition

$$N_B = \sum_{\tau=1}^{T} \left[NS_\tau(T) - NG_\tau(t_\tau) - NK_\tau \right], \tag{2.2.11}$$

where NK_τ is the number of barnacles of the τ-th age bracket dying of intraspecific competition, which can be determined in a fairly similar way from the values of the critical (or maximally possible) number NC of barnacles, which is determined from field data by the equation

$$NC = \left[N_0^{(3)} \cdot \omega_B^{-\gamma_9} \right] \cdot a_2, \tag{2.2.12}$$

where $N_0^{(3)}$ is the initial critical number averaged over options; γ_9 is a parameter representing the intraspecific competition; a_2 is a parameter for adjusting the formula for the option, Table 4. $N_0^{(3)} = 1.700$ and $\gamma_9 = 0.49$. NS_τ was assumed to be distributed uniformly over the time (over a ten-day period with an average $\overline{NS}$) and the smallest barnacles to die first of intraspecific competition. Now we have

$$NK_\tau = \begin{cases} 0 & N_B \le NC, \\ N_B - NC & N_B > NC. \end{cases} \tag{2.2.13}$$

According to (2.2.3) - (2.2.13), the food I_B assimilated by the population is

$$I_B = \sum_{i=1}^{4} e_i \cdot N_B, \qquad (2.2.14)$$

where e_i is an element of the energy budget.

The Simulation Program

The simulation program consisted of five basic and two auxiliary modules. The former were: option adjustment for uncrowded and crowded settlements (M1), adjustment of a multi-layer population version (M2), of an uncrowded settlement (M3), of a crowded settlement (M4), and a multi-layer settlement (M5); the auxiliary modules were of statistics and interaction (ouput to) peripherals (M6) and print-out and plotting (M7). The total operating time of the program was 4.30 minutes, the net CPU time 4.12 minutes, and the operation without print-out 3.07 minutes. The memory needs are shown in Table 5a.

Three common locations were employed to save the editing time and simplify the inter-modular links. Each stored parameter has values for models of one of three settlement types. The constants were introduced into the modules by formal parameters of procedures. The maximal time modelled in one run was 65 days.

The Simulation Design

For the investigation and quantitative estimation of the basic characteristics of an energy flow through the balanus population, three series (options) of simulation, with 12 runs each, were performed. Each series applied to a certain type of settlement and period in the life. Abiotic factors varied from series to series because, for instance, the salinity dictates the type of settlement and the impact of this type on the hydroid.

The runs in each series differed in the number of phytoplankton cells (α) in thousands per cubic meter. Besides, for an uncrowded settlement, runs were carried out for various numbers of continuously living specimens, Table 5b. The range of α was 10 to 300. The simulation field data led to the equation

$$CW_B(\alpha) = a_3 \cdot \alpha^{b_3}. \qquad (2.2.15)$$

Table 5a Memory needs of the simulation program for the balanus population model

User	Number of FORTRAN statements	Core memory in Kbytes
M1	61	18
M2	33	22
M3	38	2
M4	53	3
M5	85	5
M6	41	3
M7	120	6
Editing	-	108
Execution	-	98
Entire program (initial module)	451	59

For an uncrowded settlement (2.2.6) is replaced by the equation

$$A_B = a_4(\alpha) \cdot \omega_B^{b_4(\alpha)} \quad (2.2.16)$$

with

$$a_4(\alpha) = a_5 \cdot \alpha^{b_5} \quad (2.2.17)$$

and

$$b_4(\alpha) = a_6 \cdot \alpha^{b_6}. \quad (2.2.18)$$

The values of parameters in the equations (2.2.14) - (2.2.18) are given in Table 6.

The coefficients of Table 6 are a sort of convoluted information on the population as a function of food concentration, information obtained by simulation with the method of least squares. As the resulting regression is different from the simulation experiment data for all the dependences by a maximum of at most two to three per cent, the significance of these factors may be equal to that of the initial equations in Section 2.5.

As in all the other cases, the adequacy was analyzed by ascertaining the consistency of findings by field experimental data omitted in the research, and by qualitatively fitting the computed values with the ecology of the system (Turpayeva and Simkina, 1975; Turpayeva and Gal'perin, 1980 a & b). In the field experiment with man-made plates the food concentration was 180,000 specimens/m^3 water. The discussion below will refer to this version. The results on the multi-layer balanus settlement are most relevant for the balanus - hydroid relation. The field data for this type of settlement were obtained in plates where some hydroids were observed. The numerical characteristics will be discussed in Chapter 3 below.

Table 5b Values of ω_0 (with $\omega_B > \widetilde{\omega}_g$) in equation (2.2.1) and the number (N_B) of specimens developing in an uncrowded settlement

Value (thousand) / Parameter	10	30	50	60	70	100	120	150	180	200	250	300
ω_0	6.04	5.26	4.83	4.72	4.47	4.31	4.16	3.98	3.65	3.75	3.55	3.49
N_B	15	15	18	20	20	120	120	120	226	1,200	1,200	1,200

Table 6 Parameters of equations (2.2.15) - (2.2.18)

Parameter / Weight of specimen	a_3	b_3	a_5	b_5	a_6	b_6
$\omega_B \leq \omega_{max}$	0.076	0.40	-	-	-	-
$\omega_B > \omega_{max}$	0.003	0.64	0.11	0.37	0.80	0.03

For uncrowded and crowded settlements the energy of food assimilated by the barnacles and spent for growth, energy metabolism, regeneration, and moltings is shown in the Figures 11 - 13. Early in the model time, a small fraction of the assimilated food energy is spent for growth. Barnacles in uncrowded settlement spend from 66.5 to 84.5 per cent of the total energy for growth during the first 10 days and in a crowded

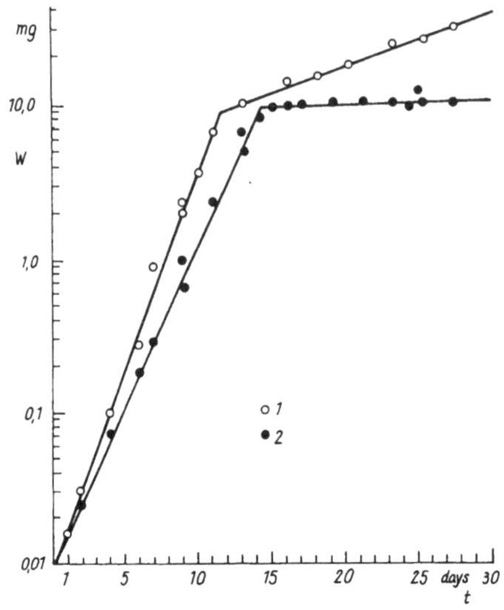

Fig. 11 Balanus growth on experimental plates:
1 - uncrowded settlement; 2 - crowded settlement.

settlement from 60.0 to 81.8 per cent during the first 13 days. After the eleventh day the growth of barnacles in an uncrowded settlement dramatically slows down and accounts for as little as 9.1 - 9.7 of the daily food energy. The balanus from a crowded settlement practically stops growing on the 14th day. The barnacles in an uncrowded settlement spend 4.1 to 5.8 per cent of the energy for moltings, and in a crowded settlement, 4.7 to 6.9 per cent. The fraction of energy for energy metabolism decreases with time, in barnacles from an uncrowded settlement from 29.4 to 10 per cent in the first ten days and to 6.5 per cent in the following 20 days; in barnacles of a crowded settlement, from 35.2 to 11.0 per cent in the first 13 days and to 10.5 per cent in the subsequent 16 days. After the 11th day in an uncrowded settlement and the 13th day in a crowded settlement, the bulk of the daily energy of assimilated food is taken up by regeneration. In the barnacles from an uncrowded settlement this percentage is 77 - 78 and in those from a crowded settlement 81 per cent.

The results obtained in this research on elements of the input - output equations for the balanus fit well the earlier findings (Sushchenya, 1967; Tamkevicene, Sestokas, 1975; Brauch, 1975; Grahame, 1973; Hughes, 1970; Kuenzler, 1961; Wilbur, 1976; Wu and Levings, 1979). The barnacles that develop in crowded settlements grow slower and weigh less. The 30 day total energy spendings for the above processes in barnacles from a crowded settlement are smaller than in those from an uncrowded settlement, Table 8. The ratio of the energy amounts spent by barnacles from

Table 7

Module	Basic equations or activities	Level in mathematical flowchart	Level in structural flowchart
M1	Adjustment for the settlement feature, specifying food concentration, salinity, generation of common memory locations obtaining parameters from library, calling simulation modules; print-out.	IV	V
M2	Input of version feature, input of data on settlement by ten-day periods, reading option parameters, specifying parameters of common memory locations and simulation cycles; print-out.	IV	V
M3	Simulation of population development in an uncrowded settlement by the equations (2.2.1), (2.2.3) - (2.2.7); print-out.	I, II	I - IV
M4	Simulation of population development in a crowded settlement with equations (2.2.1), (2.2.3), (2.2.8); print-out.	I, II	I - IV
M5	Simulation of population development in a multi-layer settlement by the equations (2.2.9) - (2.2.17); print-out, whenever necessary continued simulation, goto M2.	I, II	I - IV
M6	Computation of auxiliary statistics such as K_2, Table 8, and parameters of uniform distribution among days in a ten-day period of settlement; input/output of data and results from library.	IV	III - IV

a crowded settlement to those spent by barnacles in uncrowded settlements for growth, energy metabolism, molting, and regeneration are not the same. During thirty days barnacles from a crowded settlement spend for growth one forth, for metabolism five ninths, for molting 10/23, and for regeneration 5/14 of the energy spent by barnacles in an uncrowded settlement for those purposes. This difference is evidence of the different contributions made by different energy processes to the overall life strategy of the balanus. Using the relations (2.2.1) -

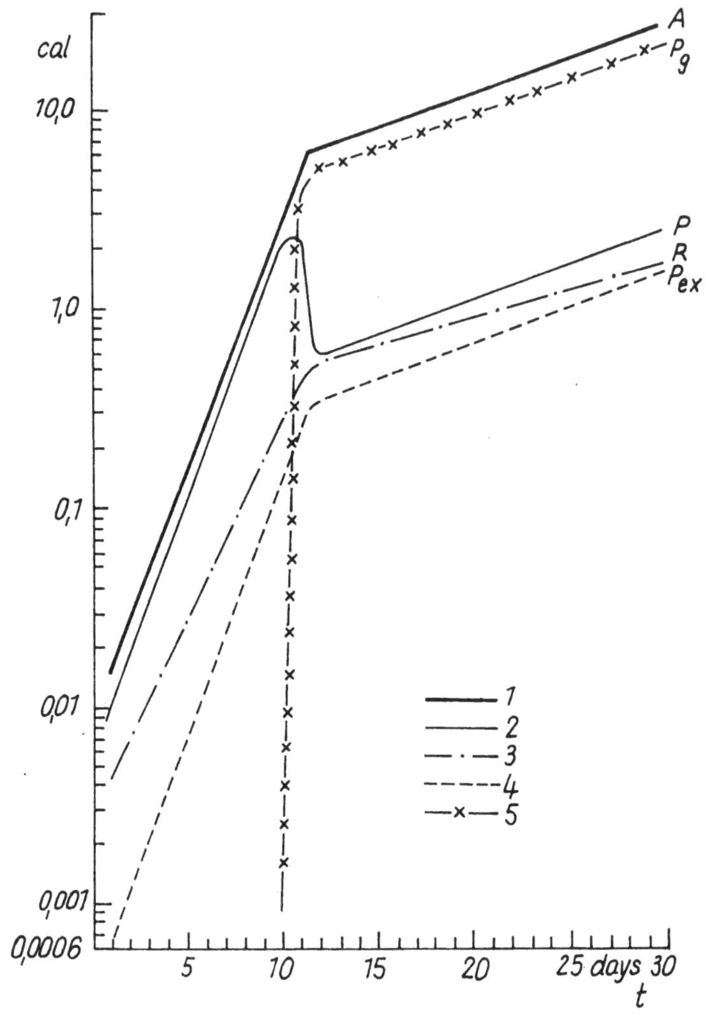

Fig. 12 Elements of energy balance versus time for the balanus in an uncrowded settlement (Turpayeva, Gal'perin, 1980 a): 1 - assimilated food; 2 - spent for growth, 3 - energy metabolism; 4 - moltings and 5 - regeneration.

(2.2.8), for every day of the model time, numbers and biomasses of the balanus were computed for a crowded single-layer settlement. The adequacy of the results was checked by comparison with data for a single-layer fouling in experimental plates at the input of water-intake installations in 1969, Fig. 14 (Turpayeva and Gal'perin, 1980 b).

The energy flow through a modelled balanus settlement in one day increases with time. In an uncrowded settlement the energy flow increases from 0.003 to 1.0 kcal/day during the period of a high growth rate (the first 11 days of the model time), and during regeneration (the subse-

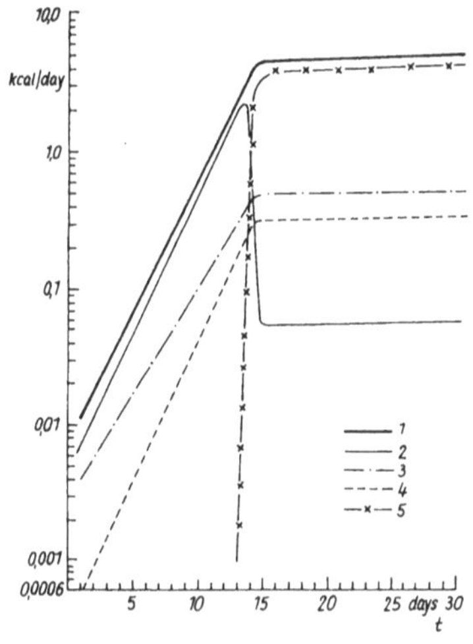

Fig. 13 Energy balance of elements versus time for the balance in a crowded settlement. The notation is as in Fig. 12 (Turpayeva, Gal'perin, 1980 a).

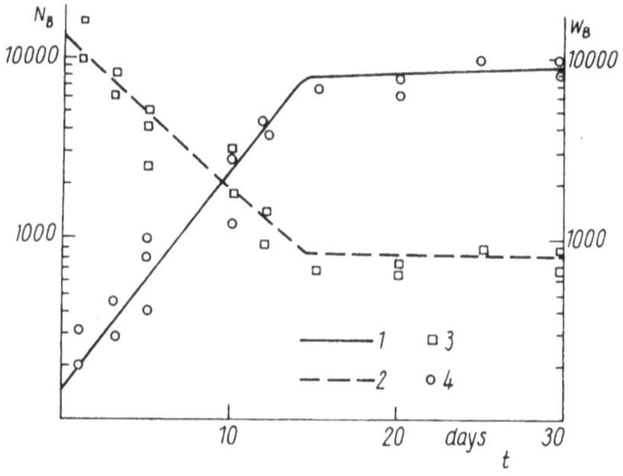

Fig. 14 Characteristics of a crowded balanus population versus time: 1 & 2 - computed curves; 3 & 4 - empirical data; 1 & 4 - biomass, dry weight, mg/dm^2; 2 & 3 - number of barnacles per square decimeter; N_B - number, specimens.

quent 18 days) it is as high as 5.5 kcal/day. In a crowded settlement the energy flow increases from 0.14 to 2.7 kcal/day in the first 14 days (growth) and to 3.2 kcal/day during regeneration, Fig. 15.

The energy allocation for life processes changes with time in both model settlements, Fig. 16. Initially most of the daily energy flow is taken up by growth and then by regeneration. During the first 10 days in an uncrowded settlement and the first 14 days in a crowded settlement the fraction of energy for growth gradually increases from 63 to 84 and from 59 to 82 per cent, respectively, and the percentage of

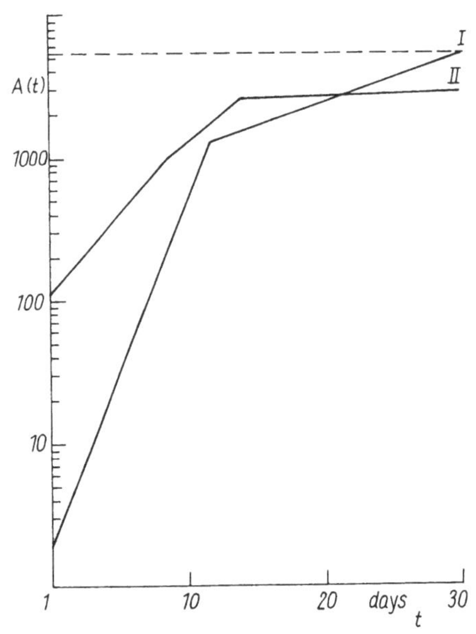

Fig. 15 Energy flow through model balanus settlements (A_B, kcal/day) (Turpayeva, Gal'perin, 1980 b):
I - uncrowded settlement; II - crowded settlement.

energy used in energy metabolism decreases from 29 to 10 and from 35 to 11 per cent, respectively. Regeneration in an uncrowded settlement accounts for about 77 per cent and in a crowded settlement, for 82 per cent of the daily energy flow. In a crowded settlement, a major part of the daily flow, 18 to 25 per cent, is taken up by barnacles that later die.

Towards the end of the model time, the biomass of a crowded settlement was two thirds and the number twice that in an uncrowded settlement, Table 8a. The crowded settlement spent ten per cent more for growth, energy metabolism, and molting, while, for regeneration, ten per cent less than the uncrowded settlement did.

The numerical simulation of settlement development revealed that most of the energy of assimilated food is taken up by regeneration, and the smallest fraction by moltings. The energy metabolism and somatic growth spendings were approximately equal. The net efficiency of the overall balanus growth was very high, for growth (with molting and regeneration) accounted for about 90 per cent of the energy of assimilated food. The population density (in the ranges used in the computations) has almost no impact on these indicators of model populations (Turpayeva and Gal'perin, 1980 a & b).

The population density has a noticeable effect on the quantitative development of settlements and the daily energy flow through them. The range of the daily flow versus time is much wider for an uncrowded

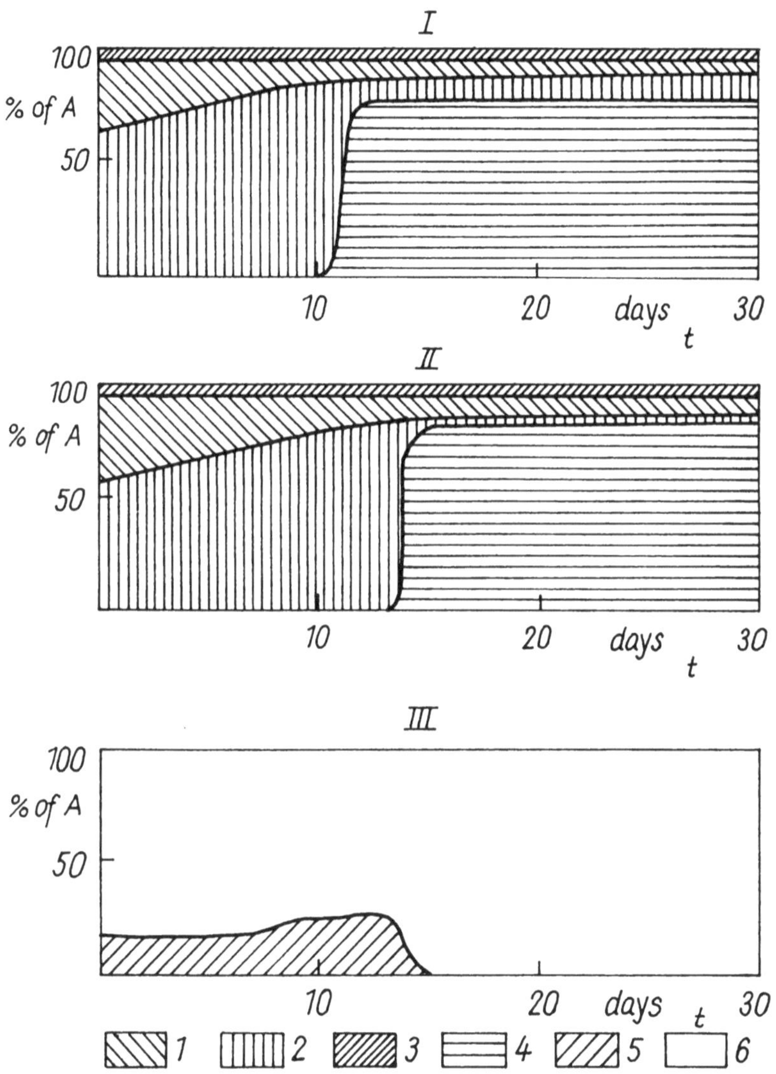

Fig. 16 Energy flow allocation for the balanus life processes (Turpayeva, Gal'perin, 1980 b):
I - uncrowded settlement; II & III - crowded settlement. Fraction of energy spent for: 1 - energy metabolism, 2 - growth, 3 - moltings, 4 - regeneration, 5 - development of barnacles bound to die, and 6 - activities of surviving barnacles.

settlement, but, with time, the differences are offset and the total energy flow through the model population during its lifetime is nearly the same. Since, in a crowded settlement, some of the flow gets lost due to elimation, in a natural settlement it is probably energetically more advantageous for the species to form uncrowded settlements taking up the entire substrate surface. Such settlements have a larger biomass and produce more larvae, but are less stable in the face of exogenous impacts. The death of some specimens in such a settlement would inevitably reduce its productivity. A crowded balanus settlement may prove to be more stable in the face of unfavourable exogenous events because relatively weak specimens are eliminated. The relatively low energy consumption for the development of barnacles which are bound to die and some reduction of the general fertility may be considered the price to be paid for the increased stability of the settlement.

Consequently, the consistency of the computed results and the field experiment, Figures 11 and 14, may be one of the proofs that the balanus population model adequately represents reality. The consistency of the qualitative results of simulation and the general ecological facts known about fouling species, notably the balanus, seems to confirm the significance of the numerical values obtained in modelling. These values may be found useful in forecasting the development of the balanus in various kinds of settlements on man-made surfaces, if the actual system consists mainly of the monoculture of the species the research is concerned with. This situation is quite realistic (Turpayeva, 1967). What is more, simulation has revealed specific features of the system. One of these is the high rate of energy flow through the population. In the lifetime of just one spring generation the energy flow through a balanus settlement is commensurable with that through populations of mobile bottom invertebrates within a year (Turpayeva, Gal'perin, 1980 a & b; Holm and McIntire, 1971).

Table 8a Energy flow through model balanus populations over 30 days (per 1 dm^2)

Settlements				Uncrowded	Crowded
Number				226	555
Biomass, g				9.09	6.07
Assimilated food				59.97	61.32
Food energy allocation	kcal	% of A_B	kcal	% of A_B	
Energy metabolism	4.42	7.4	7.25	11.8	
Production P_B	6.99	11.7	9.84	16.0	
Production PEX_B	3.40	5.6	4.13	6.7	
Production G_B	45.16	75.3	40.11	65.4	

Table 8b The total balanus energy budget over 30 days in model settlements

Population	ω_B	Energy budget elements					$K_2' = \frac{P_B + PEX_B + G_B}{A_B}$
		A_B	P_B	R_B	PEX_B	G_B	
Uncrowded, cal	30.94	265.34	30.93	19.57	15.03	199.81	0.93
% of A_B	-	-	11.7	7.4	5.6	75.4	
Crowded, cal	7.83	97.08	7.82	10.61	6.63	72.03	0.89
% of A_B	-	-	7.9	11.0	6.8	71.3	

2.3. Nudibranch and Infusiroan Population Models

2.3.1. The Nudibranch Model. The Object

The nudibranch Tenellia adspera nudibranchia, like the hydroid and balanus, is a mass species in the fouling biocoenosis on water-intake installations. Tiny mobile nudibranches subsist on hydroid colonies, whose growth they slow down, and change the contributions made by elements to the energy budget in the total assimilated food. In winter, the nudibranch is not present in the colony. Its individuals appear in late May and early June, and in July there are tens of thousands of nudibranches per 100 m^2 in some parts of the hydroid colony. The nudibranch is hermaphroditic and breeds by attaching eggs to the substrate. For an unlimited population the dynamics is dictated by the following processes:

1. Fertility variations with the age of the specimen.
2. Maturity onset for different age groups.
3. Reduced number of eggs as the juveniles appear.
4. Death of aged specimens.

The per diem fertility of growing animals varies with time and depends on the water salinity and age of the specimen and is at a maximum when the animal reaches its maximum size. The population at the begin consists of different numbers of adult and juvenile specimen. In different options two, four or eight mollusks per square meter of the hydroid fouling were observed.

Formalization

The first version of the nudibranch population for unlimited development (Turpayeva and Lebedeva, 1976) formalized the dynamics of the population in a survival matrix for a fixed eight per mille salinity; the energy flow through the population was computed from the energy input - output of one specimen.

To apply the nudibranch population model in simulation studies of the fouling community, the mathematical description had to be adapted to the approach of this research, and new data on the nudibranch population had to be incorporated.

The linear specimen growth l can be described as a power time function in the light of available field data

$$l = c_4 \cdot t^{c_5}. \qquad (2.3.1)$$

The parameters of this equation are considered to be constant for two salinity levels, five and eight per mille, Table 9. The nudibranch fertility is computed from field data by the method of least squares for the five per cent significance level of the logarithmic regression.

$$n_T = c_6 \cdot t^{c_7}. \qquad (2.3.2)$$

The weigth and the energy value of the nudibranch specimen are directly proportional, the factor having been obtained experimentally (Turpayeva and Simkina, 1975; Turpayeva and Lebedeva, 1976). Taking into account the results on one specimen and the nudibranch's sensitivity to salinity variations, the assimilated food of one nudibranch specimen as a function of its weight is

$$a_T = c_8 \cdot \omega_T^{c_9}. \qquad (2.3.3)$$

The numerical values of the parameters c_{4-9} are shown in Table 9, which also summarizes the data on the development and survivability of nudibranch larvae and juveniles for two salinity levels.

The mathematical description of the population dynamics also called for some updating in compliance with the structural diagrams and salinity data. The number of specimens in the τ-th generation can be represented, according to (2.3.2), by the equation

$$N_\tau = \sum_{\tau = T - t_p}^{T-1} N_{\tau - 1} \cdot n_T (T - \tau - t_p^*), \qquad (2.3.4)$$

where t_p = 20 (the specimen's lifetime) and t_p^* is the mature age. Taking into account the survivability, this equation turns into

$$N_\tau = \sum_{\tau = T - t_p}^{T-1} \left[N_{\tau - 1} \cdot V_M(S)\right] \cdot \left[n_T(T - \tau - t_p^*) \cdot V_L(S)\right]. \qquad (2.3.5)$$

Now the total number of living nudibranches can be expressed by the equation

$$N_\Sigma = \sum_{\tau = T - t_p}^{T} N_\tau. \qquad (2.3.6)$$

In the light of the above equations and with due regard for the elements e_i of the specimen's energy budget the nudibranch's population energy flow is represented by the following equation

$$A_T = \sum_{i=1}^{3} \left[\sum_{\tau = T - t_p}^{T} (e_i (t_\tau) \cdot N_\tau) \right] \qquad (2.3.7)$$

where t_τ is the age of the τ-th generation.

Characteristics of the Simulation Program

The program which simulated the activities of the nudibranch population consisted of three basic and two housekeeping modules: The former included option adjustment and statistics (M1); computing the elements of the specimen's energy balance (M2); and computing the size and characteristics of the population (M3); the latter were interaction with external media (M4) and print out and plotting (M5). The total run time at the program was 3.1 minutes, the net CPU time 1.09 minutes, and the program run without printout, 2.18 minutes. The memory needs are presented in Table 10.

Table 9 Numerical values of parameters in equations for the nudibranch specimen's balance and survivability data

Parameter / Salinity S, o/oo	C_4	C_5	C_6	C_7	C_8	C_9	Larvae survivability %, $V_L(S)$	Juvenile survivability %, $V_M(S)$
5	0.31	1.18	0.00025	5.07	1.99	0.91	87	90
8 - 11	0.33	1.21	0.00049	5.45	3.48	1.03	98	97

Table 10 Memory needs for the simulation of the nudibranch population

Characteristic / User	Number of FORTRAN statements	Core memory, Kbytes
M1	38	22
M2	40	1.5
M3	74	1.9
M4	29	1
M5	120	6
Editing	-	112
Execution	-	68
Total program (initial module)	301	32.4

Simulation Schedule

The numerical values of basic characteristics of a nudibranch population under no constraints were obtained in two series of simulation experiments which differed in the parametrically specified salinity. The equations (2.3.1) - (2.3.4) show that the numerical values should be noticeably different in the two series because the equation parameters depend on salinity. In either series, three versions were different in the assumed initial number of regenerating specimens (NR_0) per unit area at model time T = 1: (a) $NR_0 = 2$; (b) $NR_0 = 4$; and (c) $NR_0 = 8$. Like salinity, these values were specified parametrically.

Table 11

Module	Main equations or operations	Level in mathematical flowchart	Level in structural flowchart
M1	Series adjustment, storing parameter values in common memory areas; version adjustment for the initial number; data access from library; call of simulation modules; print-out; standard statistics.	IV	V
M2	Simulation of specimen growth, computing energy balance by the equations (2.3.1) - (2.3.4); print-out.	I	I, II
M3	Computing general population characteristics from population size simulation results by the equations (2.3.5) - (2.3.7); print-out.	II	II - IV

Model Adequacy and Some Numerical Results

The numerical results on simulating the development of a population in an unconstrained medium for two salinity levels and an initial nudibranch population of two specimens are summarized in the Figures 17 and 18. Figure 17 also presents field data (Turpayeva and Levedeva, 1976). The complex age structure of the population and the different numbers in adult generations result in sine-wave population variations. As the number of multiplying generations increases, so does the wave period. The population structure changes with time. In the initial period

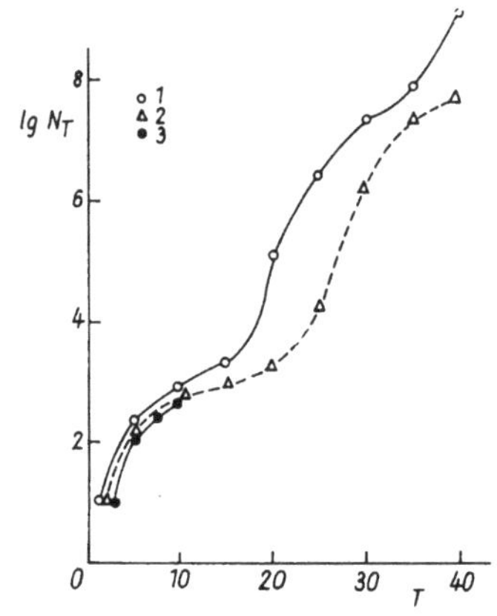

Fig. 17 The nudibranch population dynamics for different salinity levels: 1 - 8 o/oo; 2 - 5 o/oo; 1 & 2 is calculated and 3 are field data.

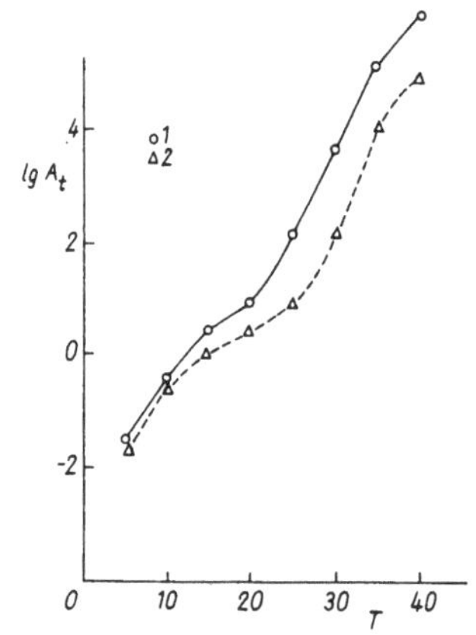

Fig. 18 The energy dynamics of food assimilated by the nudibranch for salinities:
1 - of 8 o/oo; 2 - of 5 o/oo.

juveniles are prominent, but by the 40th day adult nudibranch account for about 3.2 per cent of the total. The kind of assimilated food on the whole was consistent with the population dynamics. The difference in the period and amplitude of the food oscillations corresponds to the gradual increase in the number of adults. The least variable characteristic was the metabolic energy in the total assimilated food energy that ranged from 25 to 35 per cent over the entire modelled interval. When there were less than five per cent of adults in the population, most energy was taken up by the growth. When that number increased, the fraction of regeneration energy increased dramatically. Over a long time interval the population dynamics in terms of totals fits well the data of ecological research (Turpayeva, Simkina, 1975). Furthermore, the population dynamics over 10 days fits well the control data. For this reason the model may be viewed as adequate.

In simulating the population dynamics with a five per cent salinity, the population increase slowed down. The numbers varied less but at a higher rate, which corresponds to the higher fertility at the five per mille salinity. By the 30th day the assimilated food of the population with low salinity is approximately one eightieth in comparison with that with eight per mille. Moreover, the general pattern of the processes in a population is largely the same in the two series. Table 12 presents some numerical results on the series with an initial population of two specimens.

Table 12 Energy value of the weight (ωK_T), the assimilated food (A_T), and the number of mollusks (N_T)

Quantity	S = 5 o/oo			S = 8 o/oo		
Age days	ωK_T (kcal)	N_T	A_T (kcal)	ωK_T (kcal)	N_T	A_T (kcal)
20	0.58	$6.29 \cdot 10^3$	0.34	9.25	$11 \cdot 10^3$	6.01
30	$1.2 \cdot 10^2$	$1.4 \cdot 10^6$	$0.7 \cdot 10^2$	$9.1 \cdot 10^3$	$106 \cdot 10^6$	$6.0 \cdot 10^3$
40	$25 \cdot 10^3$	$289 \cdot 10^6$	$14.4 \cdot 10^3$	$8.9 \cdot 10^6$	$103 \cdot 10^9$	$5.9 \cdot 10^6$

The simulation results for initial populations of four and eight did not fit the field data and so only the above case was considered.

2.3.3. The Infusorian Population Model

The Object to be Modelled

The colonial infusorian Zoothamnium hertscheli Kahl develops on the generative part of the hydroid colony and is the food of the fouling community. If its substrate is not constrained, the infusorian colony grows exponentially. Field studies of this population revealed the effect, that, if the generative part of the hydroid is fouling, the infusorian population tends to cease growing. In a field experiment the infusorian population reached a maximum in four to six days. For simulation the number and weight variations were important, while the energy balance components were not estimated.

Formalization

The maximal infusorian population per square centimeter of hydroid stolons was estimated in a field experiment to average 3,100 to 3,200. The exponential increase with a dramatic slowing down of the growth when the substrate area was occupied led to the assumption that the population curve was S-shaped. Using the standard linearizing transformation which leads to linear pairwise regression the factors were obtained for a logistic equation

$$N_S = N_{ms}/(1 + \exp(c_{10} + c_{11} \cdot t)), \qquad (2.3.8)$$

where N_{ms} = 3,200 specimens/cm^2, C_{10} = 2.9, and C_{11} = −1.1. Unfortunately, there were only three test points for the infusorian population size; as a result, the significance of the regression equation proved to be low. The energy value of the colony per square centimeter of the hydroid stolons is expressed by the relation

$$\omega k_s = \omega_s \cdot \Psi_s \cdot N_s, \qquad (2.3.9)$$

where ω_s is the weight of one infusorian and Ψ_s is its energy value ($1.674 \cdot 10^{-7}$ and $5.05 \cdot 10^{-8}$, respectively). With the data for an area of one square centimeter known, that for the entire or part of the generative area of the hydroid colony is

$$S_p = W_{\Sigma}^{s} \cdot K_G \cdot N_0/L_p, \qquad (2.3.10)$$

where W_{Σ}^{s} is the substrate weight; $N_0 = 865 \cdot 10^4$ is the number of hydranths in a colony whose weight is 1 kg; K_G is the hydranth's energy value, $0.7 \cdot 10^{-5}$ kcal; and $L_p = 0.55 \cdot 10^{-4}$ kcal is the energy value of one centimeter of stolons. The diameter of the cylindrical stolons averages 0.02 cm.

The Simulation Program

The simulation program of the infusorian population model consisted of three modules: initial conditions and printing; population computing; and output to an external medium. The program consisted of 68 FORTRAN statements and needs one Kbyte of core memory. The overall execution time was two minutes and every execution step needed 50 Kbytes of core memory.

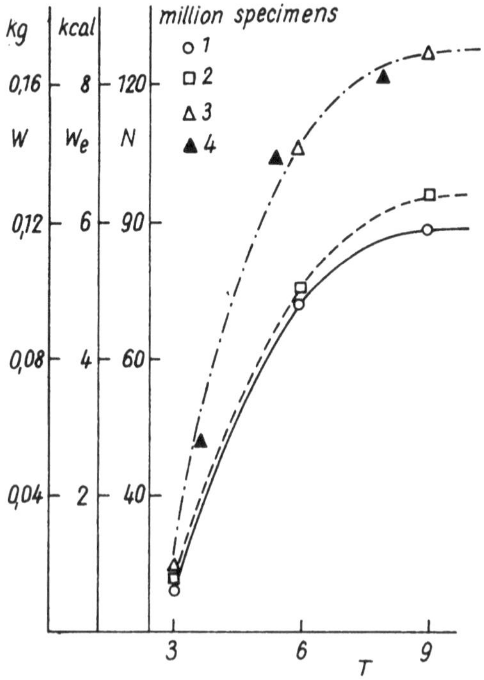

Fig. 19 The dynamics of the infusorian population:
1 - computed; 4 - field data; 2 & 3 - computed dynamics of the population energy value and biomass.

The Simulation Design

In computing the population size and energy value in various substrate areas, the latter were determined by the equation (2.3.10), which is consistent with various ecological findings. In particular, in terms of the biomass of the generative part the relation (2.3.10) for high concentrations of the hydroid food yields about 500,000,000 infusoria. Field observations suggest that the parts of the hydroid colony where the infusorian develops account for 20 to 40 per cent of the total polyp biomass.

Model Adequacy and Some Numerical Results

Figure 19 compares the computed and field data for a controlled substrate area. The data per kilogram of the hydroid polyp show that the maximum of the infusorian population is achieved in four to six days and is 133,000,000. The data of infusorian development are more significant when interspecific relations are admitted. Therefore, the dynamics of this population will be discussed in more detail in Chapter 3, where the impact of the hydroid substrate and various interspecific relations on the infusorian will be analyzed.

2.4. The Crab Population Model

The Object

Like the other populations described in this book, the crab Rhithropanopeus Harrisii tridentatis is widely spread in the fouling communities of the Sea of Azov. The dynamics of the crab population and its energy characteristics were studied in the vicinity of Zhdanov in the summer and fall of 1971 (Turpayeva, 1972; Gal'perin et al., 1981). The hydroid substrate for the crab is a two-layer fouling 10 to 20 cm thick and having a biomass of about 8 kg/m^2, the upper layer consisting of growing and regenerating offshoots (the regenerating part of the colony), and the lower one (the somatic part) consisting of intertwined hydranthless perennial stolons (Turpayeva, Gal'perin, and Simkina, 1977). Crabs are to be found in both layers, but mostly in the upper one. In the smallest crabs the carapace diameter is 1.4 mm and in the largest ones, 20 mm. The crabs were classified into eight groups according to the number of moltings of crabs that arrived in the current summer. The average values in the classes were 1.8, 2.4, 3.2, 4.2, 5.5, 6.9, 9.4, and 12.7 mm.

In the winter and spring the crabs whose carapace diameter exceeds 5 mm were prevalent (Turpayeva, 1972). The juveniles appeared in summer and stood in the fouling until late in the fall. In 1971 the crab larvae, megalops, started to settle on a massive scale on July 4. Since that time juveniles whose number varied with time, were always present. Let us visualize the processes which dictate the number of crabs in the fouling community (Gal'perin et al., 1981).

1. Megalop settling, molting, and transformation into juveniles. This stage was omitted in the model because its duration was uncertain.
2. Growth of specimens, Table 13.
3. Reduction of the number of crabs as they grow, with some of them dying or being ousted from the substrate.
4. Migration inside the substrate and between the strata.

The total number (N_Σ) of crabs and their number (n_L) in each of the eight (L) classes of size can be obtained by adding the number of juveniles joining the population every day and the number of large crabs (n_{FL}) which have survived the winter. The variation in the group of crabs which arrive in one day (to be referred to as a τ-th age group) is due to natural deaths, consumption by large crabs, migration inside the community and, once these crabs (n_τ) grow to large sizes, death due to territorial behaviour understood as migration beyond the hydroid substrate area of large crabs as a result of stiffer intraspecific competition. This competition seems to be caused by trophic as well as topic factors. Since they could not be separated in the field or in the mathematical model, these processes will be considered as one process in the following.

Formalization

Arranging the initial data according to the size and weight made it possible to identify the variation in the number of crabs (n_τ) in different age groups. Different kinds of dynamics were discovered (a) when the lower parts of the hydroid fouling were not taken up and the juveniles could migrate there; (b) when both the lower and upper parts were not taken up; (c) when both parts were partly taken up; and (d) when the lower part was not taken up and the upper was by medium and large crabs, which hindered the reemigration of grown crabs.

Table 13 Size, age, and inter-molting intervals of the crab (Turpayeva et al., 1981)

Sequential number of inter-molting period (L)	Crab size (d) mm	Duration of inter-molting period (t_p), days	Crab age at the time of molting (t_d), days
1	1.8	4	1
2	2.4	6	4
3	3.2	8	10
4	4.2	9	18
5	5.5	11	27
6	6.9	20	38
7	9.4	30	58
8	12.7	51	88

Important is that the biological approach reveals that the control groups, which enjoy the most favourable conditions because they are separated from the remainder of the community, have a number practically invariant until natural death. This suggests that the above intrapopulation and interspecific processes result in significant differences in the dynamics of different age groups. The expansion of the part of the Table on initial numbers in that approach led to a general evaluation of the processes (selecting the pattern and parameters in the relations) which lead to the death of the crabs, like as elimination (E_τ) and migration inside the community (M_τ). Following the research traditions, these integral processes were assumed to be described by power dependences. The part of the Table (matrix) which was concerned with the initial number and dynamics (data for several days in a period of three months) was not used in selecting the equation parameters so as to use it in adequacy analysis (Gal'perin et al., 1981).

Now, the equation for the size of one age group is

$$n_\tau(t) = \mu_\tau \cdot E_\tau(t), \qquad (2.4.1)$$

where μ_τ is constant (the number of settled juveniles) and $E_\tau(t)$ is the elimination factor. Until the day t^*, when the crabs of the τ-th age group grow to the average size, μ_τ is equal to the number of crabs a_τ of the first age group ($L = 1$), which join the population on the τ-th day. For the remaining time interval t the value of μ_τ is computed by a power dependence whose basis is the number of crabs on the day t^*. The analysis of empirical data led to the assumption that the juveniles join the population daily during the entire model time. The size of the population a_τ was computed for every day of the model time by one of the auxiliary procedures of the simulation program. Its parameters were the time during which the entire population existed, the number of all juveniles ($L = 1$), and the value of factors of total losses in the life of these crabs before the first molting.

The elimination factor E_τ was computed as a product of two functions, $E_\tau(t) = f_1(t) \cdot f_2(t)$. One of these functions, f_1, was assumed to describe the processes of natural death and the consumption of the juveniles by larger animals, and the other, f_2, the death of adult crabs corresponding to territorial behaviour. In the model, both are exponential - in power dependence on the age t_τ. In the computation, the impact of elimination was recognized for the total number of crabs in a given age group in the lower as well as the upper layer of the fouling. In order to allow for the migration from the upper layer to the lower one, a factor M_τ was introduced and quantitatively estimated. Using (2.4.1) and denoting $g_\tau = 1 - E_\tau(t)$ and $m_\tau = 1 - M_\tau(t)$, the number of crabs in every age group in the upper layer, n'_τ, was determined as

$$n'_{\tau}(t) = \mu_{\tau} \cdot \left[1 - (g_{\tau} + m_{\tau}) + g_{\tau} \cdot m_{\tau}\right], \qquad (2.4.2)$$

and in the lower level, n''_{τ}, as

$$n''_{\tau}(t) = \mu_{\tau} \cdot m_{\tau} \cdot (1 - g_{\tau}). \qquad (2.4.3)$$

Both, $E_{\tau} = 1$ and $M_{\tau} = 1$ implied the absence of the corresponding process.

Equations for the number of crabs in the population prior to mass settlement of the juveniles, $n_F(t)$, were kindred with (2.4.2). In computing n_F of every class size, the constants $\hat{\mu}_L$ (specimens) and the size wave period Z_L (days) in the migration factor were determined from the empirical data and were assumed to be $\hat{\mu}_4 = 10$, $\hat{\mu}_5 = 10$, $\hat{\mu}_6 = 3$, $\hat{\mu}_7 = 2$, $\hat{\mu}_8 = 1$ and $Z_4 = 11$, $Z_5 = 20$, $Z_6 = 10$, $Z_7 = 20$, $Z_8 = 20$ in the final computation.

The simulation revealed that, depending on τ (the settlement day), the parameters of equations describing the dynamics of the number of crabs in one age group varied in a wide range.

The numerical values of the parameters in the system of equations of the model are represented in Table 14. The heading shows the general form of the equations and time intervals during which they hold for the factors E_{τ} and M_{τ}, the parameter C, and the constant (μ_{τ}). The time intervals of juvenile settlement are shown in the second column of the Table. For those age groups which do not migrate much the totality of parameters was used, which characterizes, in Table 14, the fourth group of time intervals of settlement. For these crabs, with a weak impact of the migration factor (M_{τ}), the elimination factor (E_{τ}) was found to be highly significant because of two shortliving but high peaks of juvenile settlement and, consequently, fast filling of the lower hydroid layer. The parameters of the second and third time intervals of settlement were applied to lively migrating crabs (early in intensive and during weak juvenile settlement).

Numerical values for the parameters of the system of equations (2.4.1) - (2.4.3) were selected by an iterative simulation program (an auxiliary module of the main program), which verified the results by introducing the "setpoint" (Brandt, 1970; Afifi, Eisen, 1979).

The total number of crabs in the community is

$$N_{\Sigma}(T) = \sum_{\tau=1}^{T} n_{\tau}(t_{\tau}) + \sum_{L=1}^{LS} n_{FL}(T), \qquad (2.4.4)$$

where N_{Σ} is the total number of crabs in the community at model time T; $n_{\tau}(t_{\tau})$ is the number of crabs that joined the community on the τ-th day (τ is also the sequential number of the age group); $n_{FL}(T)$ is the number of crabs of size L which were in the community before the mass larvae settlement began; t_τ is the model time of the crab age in a τ-th age group; and L is the index of the size group (LS = 8). The number of crabs of every size was computed by a logical-heuristic program procedure with the use of data on the growth of crabs, Table 13. The equation for the number of crabs in one age group in the entire community is

$$n_L(T) = \sum_{\tau=\bar{\tau L}}^{\bar{\bar{\tau L}}} n_\tau(t_\tau) + n_{FL}(T), \qquad (2.4.5)$$

where $\bar{\tau L}$ is the sequential number of the very first τ-th age group which has size L at the current model time T and $\bar{\bar{\tau L}}$ is that of the very last τ-th age group of the same size L; t_L is the model time between moltings from size L to size L + 1 of the same τ-th age group. In (2.4.5), the number n_τ is recognized (or the time t_L changes its value) when the expressions for the upper bound BN_{up} and the lower bound BN_{dn} of the duration of crabs staying in size L have different signs. With t_p denoting the inter-molting period and t_d the crab age at the time of molting, Table 13, $\bar{\bar{BN}}_{dn} = (T - \tau + 1 - t_d)$ and $BN_{up} = (T - \tau + 1 - t_d - t_p)$. Both $\bar{\tau L}$ and $\bar{\bar{\tau L}}$ are easily computed having in mind that new crabs join the population continuously.

The expressions for time variations of one component of the energy balance BL_i in a group of crabs of size L is

$$BL_i(T) = \sum_{\tau=1}^{T} \sum_{t_\tau = BN_{dn}}^{BN_{up}} nL_\tau(t_\tau)\cdot b_i(t_\tau) + n_{FL}(T)\cdot b_i(T), \qquad (2.4.6)$$

where $nL_\tau(t_\tau)$ is the number of crabs in the τ-th age group when their size is L (L = 1, ..., 8); $h_{FL}(T)$ is the number of crabs in the community prior to mass juvenile settlement; $b_i(t_\tau)$ is the i-th component in the balance of one crab; and τ is the sequential number of the age group. The food energy AC_L accumulated by crabs of the size is

$$AC_L(T) = \sum_{i=1}^{m} BL_i(T), \qquad (2.4.7)$$

where i is the subscript of the balance element (m = 3). The total of assimilated food is

$$AC_\Sigma(T) = \sum_{L=1}^{LS} AC_L(T). \qquad (2.4.8)$$

Admitting cannibalism for the crabs, the energy flow through the population during one day, PE(T), is

Table 14

Equations and parameter values for computing the number of crabs in the τ -th age group

<table>
<tr><td colspan="2" rowspan="3">Intervals of model time (τ) of juveniles joining the population</td><td rowspan="4">t^*</td><td colspan="2">μ_τ</td><td colspan="2"></td><td colspan="5">E_τ</td><td colspan="2">C</td><td colspan="6">M_τ</td></tr>
<tr><td colspan="2">$t > t^*$ 1)</td><td colspan="2">$t \leqslant t$ 2)</td><td colspan="5">$t > t^*$ 3)</td><td colspan="2">$t > t^*$</td><td colspan="6">$t \leqslant t^*$</td></tr>
<tr><td colspan="2">$\mu_\tau = \alpha_1 \cdot n_\tau(t^*)^{\beta_1}$</td><td colspan="2">$f_1(t) = d_1 \cdot t^{-d_2}$</td><td colspan="5">$f_2(t) = d_3 \cdot t^{d_a}$</td><td colspan="2">$c = \gamma \cdot \tilde{a}_\tau^{\delta}$</td><td colspan="6">$M_\tau(t) = \tau \cdot t^{s}$</td></tr>
<tr><td>Groups of intervals</td><td>Time ds</td><td>α_1</td><td>β_1</td><td>d_1</td><td>d_2</td><td colspan="3">d_3</td><td colspan="2">d_4</td><td>γ</td><td>δ</td><td colspan="3">τ</td><td colspan="3"></td></tr>
<tr><td>1</td><td>1 - 2</td><td>1</td><td>$\tilde{a}_\tau$</td><td>0</td><td>0,983</td><td>1,212</td><td colspan="3">1</td><td colspan="2">0</td><td>1,212</td><td>0</td><td></td><td colspan="2">1</td><td colspan="3">0</td></tr>
<tr><td rowspan="2">2</td><td rowspan="2">3 - 9
32 - 40
62 - 75</td><td rowspan="2">14</td><td rowspan="2">9,512</td><td rowspan="2">0,758</td><td rowspan="2">0,991</td><td rowspan="2">0,507</td><td>$t \leqslant 25$</td><td>$25 < t \leqslant 58$</td><td>$t > 58$</td><td>$25 \geqslant t$</td><td>$t > 25$</td><td rowspan="2">0,571</td><td rowspan="2">0,053</td><td>$t \leqslant 2$</td><td>$2 < t \leqslant 11$</td><td>$t > 11$</td><td colspan="3" rowspan="2">0</td></tr>
<tr><td>0,862</td><td>1</td><td>0,071</td><td>0,301</td><td>0</td><td>1</td><td>0,01</td><td>1</td></tr>
<tr><td rowspan="2">3</td><td rowspan="2">41 - 62</td><td rowspan="2">20</td><td rowspan="2">9,521</td><td rowspan="2">0,689</td><td rowspan="2">0,994</td><td rowspan="2">0,376</td><td colspan="3" rowspan="2">1</td><td colspan="2" rowspan="2">0</td><td rowspan="2">0,381</td><td rowspan="2">0,155</td><td>$t \leqslant 8$</td><td>$8 < t \leqslant 18$</td><td>$t > 18$</td><td>$t \leqslant 8$</td><td>$8 < t \leqslant 18$</td><td>$t > 18$</td></tr>
<tr><td>1</td><td>0,242</td><td>0,172</td><td>- 0,177</td><td>0,376</td><td>0,376</td></tr>
<tr><td rowspan="2">4</td><td rowspan="2">10 - 31</td><td rowspan="2">20</td><td rowspan="2">9,508</td><td rowspan="2">0,649</td><td rowspan="2">0,211</td><td rowspan="2">0,594</td><td>58</td><td colspan="2">58</td><td colspan="2" rowspan="2">0</td><td rowspan="2">0,711</td><td rowspan="2">0,0057</td><td colspan="3" rowspan="2">1</td><td colspan="3" rowspan="2">0</td></tr>
<tr><td>1</td><td colspan="2">0,311</td></tr>
</table>

Notes: 1) For $t \leqslant t^*$ the value of $\mu_\tau = \tilde{a}_\tau$ at every τ;
2) For $t \geqslant t^*$ the parameter $d_1 = 1$ and $d_2 = C$ at every τ;
3) For $t \leqslant t^*$ the parameter $d_3 = 1$ and $d_4 = 0$ at every τ.

$$PE(T) = AC_{\Sigma}(T) - W_e'(T) \cdot U^{-1}, \qquad (2.4.9)$$

where $W_e'(T)$ is the energy value of the cannibalized crabs and U^{-1} is the consumability assumed to be 95 per cent (Chugay, Gal'perin and Turpayeva, 1981).

In determining $W_e'(T)$ it was assumed that chiefly juveniles of sizes from 1.8 to 3.2 mm fall a prey to cannibalism. $W_e'(T)$ was computed as the daily losses of the energy value of these crabs.

The Simulation Program

The simulation program for the crab population consisted of 11 main and four auxiliary modules. The former were: adjustment of the simulation program (M1); computation of additions to the population (M2); computation of background numbers (M3); computation of auxiliary statistics (M4); computing the budget for one crab (M5); one for every type of number variations (M6 - M9); computation of general characteristics (M10); and computation of total energy characteristics for the population (M11). The latter included print-out and plotting (M12); interaction with external media (M13); reduction of initial data (M14); and updating the libraries of parameters and results (M15). The total program run with the print-out is 26.41 minutes. The software flowchart is shown in Fig. 20. The modules were interfaced through two common locations of elements of the energy balance of one crab and common values for the population. The net CPU time was 5.22 minutes, and the program run without print-out 11.46 minutes. The memory requirements are summarized in Table 15.

The Simulation Design

The numerical values of the system of equations and relations which describe the model were obtained in two additional computer experiments, which were undertaken to check the assumptions about the periods of juvenile settlement and qualitative types of the dynamics describing the variations in sizes of different age groups. A dedicated program module yielded the size and weight structure of the actual number of crabs. As a result, four characteristic variations of the numbers which depended on whether the lower fouling stratum where juvenile crabs migrated was already taken up were obtained. The "background" crabs who survived the winter were found in all large sizes and their number uniformly oscillated with periods, distinct for different L. In simulating the dynamics for two and four strong short-living peaks in the juvenile settlement the model behaviour was inadequate. An additional analysis

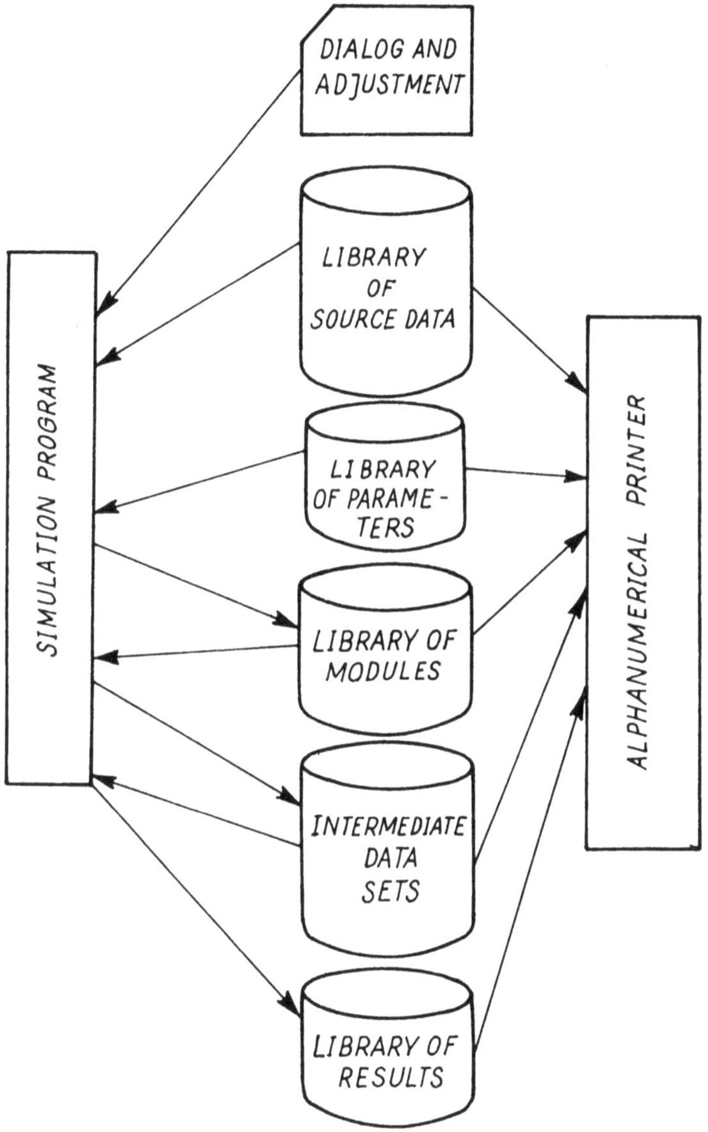

Fig. 20 Links between the simulation program and data sets in the crab population model.

of the juveniles (L = 1; d = 1.8 mm) with respect to the elimination rate and migration obtained from the preceding simulation experiment (intended to determine the types of population variation) yielded a positive result with the larvae assumed to settle continuously, an assumption confirmed later for the crab in the Sea of Azov (Turpayeva et al., 1981). In computing the budget for one crab, one experiment with stepwise and one with linear growth were performed. In the simulation model the growth was assumed stepwise (Turpayeva et al., 1981; Gal'perin et al., 1981).

Table 15

Characteristic / Consumer	Number of FORTRAN Statements	Object module core memory Kbytes
M1	63	2.9
M2	35	1.5
M3	21	1.9
M4	40	3.0
M5	94	3.1
M6	16	1.5
M7	39	3.2
M8	37	3.4
M9	26	2.2
M10	140	7.0
M11	60	2.0
M12	131	6.7
M13	21	1.7
M14	27	2.0
M15	14	1.3
Editing	-	132
Execution	-	152
Entire program (original module)	786	43.4

Model Adequacy and Some Numerical Results

For every day of the model time the simulation experiment yielded data on the crab population size. These data were plotted as the average number of crabs of every size and the number of juveniles in the upper fouling layer versus time, Fig. 21. The dots in the plots are empirical control data on the number of crabs sampled in the water intake installations. All the data, computed and empirical ones, are calculated per 100 g of undried hydroid.

The number of crabs in the model population continuously varied. In most size groups two periods of growth were noted. The two-peak pattern of the plots is dictated by the changes in the number of crabs of size 1.8 mm arriving daily, Fig. 26, curve 3. Consequently, the average of

Table 16 The functioning of main models

Model	Equations or activities	Hierarchy level in the mathematical flowchart	Hierarchy level in structural flowchart
M1	Assigning the specimen's growth type, and settlement type, reading data on initial number, Table 13; generating common memory locations; print-out.	IV	V
M2	Computing by the matrix of initial number equations (2.4.1) - (2.4.3), and Table 14 the number of juveniles settling daily; print-out; statistics of difference between control and computed numbers.	II, IV	II, IV
M3	Computing the dynamics of the "background" crab population from Table 14, matrix of initial number, and equations (2.4.1) - (2.4.3); print-out.	II	III, IV
M4	On the knowledge of the juveniles' number, additions and group totals, the dynamics type indicator is computed for the latest settled juvenile group in order to choose parameters from Table 14. The equations (2.4.1) - (2.4.3) are used to follow the dynamics of all live crabs.	II, IV	III, IV
M5	Computing the balance of one crab, reading from data library the energy characteristics versus time.	I	I, II
M6-M9	The four standard modules are parametrically independent and called and adjusted by the control module. They implement the dynamics of the appropriate type for the number of juveniles settled on the model day. Table 14 and the equations (2.4.1) - (2.4.3) are used.	I	III, IV
M10	The population energy flow is computed by the equations (2.4.6) - (2.4.8) and simulation data on the dynamics of numbers.	II, IV	IV
M11	Computing totals of the characteristics of the population size by the equations (2.4.4) and (2.4.5).	II, IV	IV

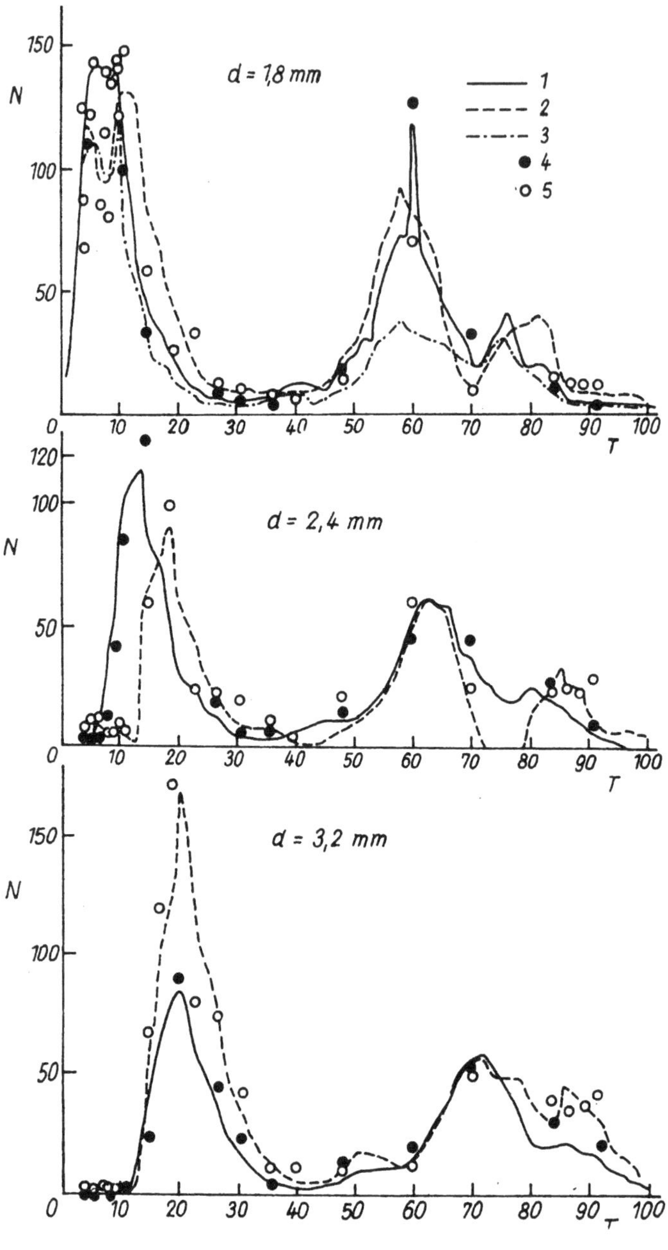

Fig. 21 The number of crabs in size groups of the model population versus time (explanation in the text) (Gal'perin et al., 1981) N - number; d - diameter; 1, 2, and 3 - computation results; 4 and 5 - experimental data; 1 and 4 - average for the entire fouling; 2 and 5 - average for the upper stratum; 3 - number of size 1.8 mm crabs settling daily.

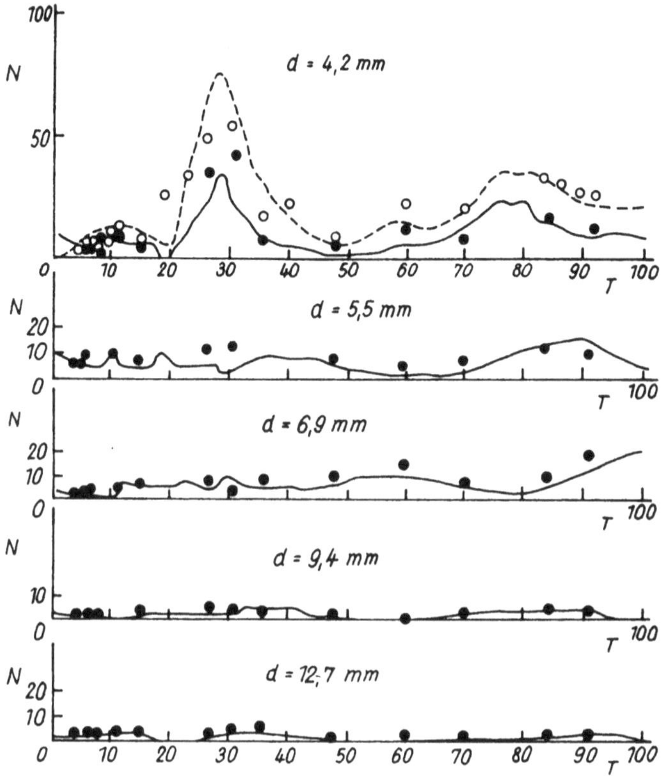

Fig. 21 (Continued)

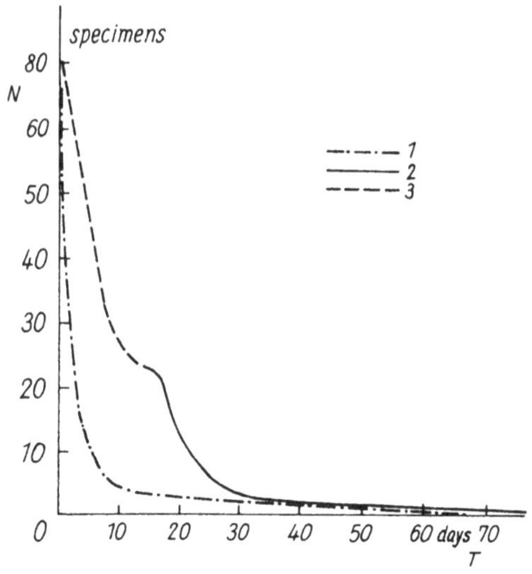

Fig. 22 **Numbers of crabs in groups in the upper (1, 2) and lower (3) hydroid strata.**

this size for the entire fouling was the highest (150 - 90 specimens) on the sixth to 11th and on the 60th day of the model time. As the size increased, the maxima of the average numbers of size groups shifted towards a later time, while their highest values decreased. The average number of crabs of size 4.8 mm reached its first maximum on the 26th to 30th day and the second one, on the 76th to 80th day (35 and 25 specimens, respectively, per 100 g of the hydroid). For crabs of size 6.9 mm the maximum (about 10 specimens) was observed on the 50th to 65th day and for size 9.4 mm the curve of the average was smoother. Juvenile crabs hid much of the time in the lower fouling layer; size 1.8 mm from the fifth to the 12th day and from the 50th to the 70th day of the model time; size 2.4 mm from the sixth to the 16th and from the 45th to the 80th day; and size 3.2 mm from the 60th to the 75th day. The number of the smallest crabs varies most dramatically with time. Thus, out of 2,025 juveniles of size 1.8 mm per 100 g of hydroid during the modelled time interval as few as 2.6 per cent remained until the last day (Gal'perin et al., 1981). This value could, however, be still lower if the number of megalops prepared for molting in this community had been reliably established. The reported data (Turpayeva, 1972) suggest that the crab population includes a maximum of 10 per cent of settled megalops. The total number of size 1.8 mm groups of crabs joining the fouling daily decreased as the population developed. This reduction was most pronounced in large groups. The behaviour of crabs in the hydroid substrate influenced the pattern of curves representing the number in the groups versus time. The extreme cases of such curves are shown in Fig. 22. The plot shows clearly that the stay in the lower stratum reduces the death rate of the juveniles.

The size-wise composition of the model population varied remarkably with time. Early in the model time, large crabs which had survived the winter were predominant in the population and the number of specimens in different size groups was nearly the same, Fig. 23. When a relatively large number of juveniles joined the community, one group outnumbered the rest. With time, the size in the dominating group increased while the percentage of such groups in the total population decreased. In particular, on the 10th day of the model time 56 per cent of the total population was made up of size 1.8 mm crabs and on the 40th day crabs of size 5.5 mm accounted for 35 per cent of the total. Less intensive but more protracted settlement of larvae in the latter half of the model time also resulted in the domination of one group (size 1.8 mm crabs accounted for 56 per cent of the total model population). As time went by, however, the domination decreased faster than in case of short-living settlement in the first half of the model time. Towards the end of the model time the percentage of juveniles was negligible, adults accumulated in the population whose composition was expected to be

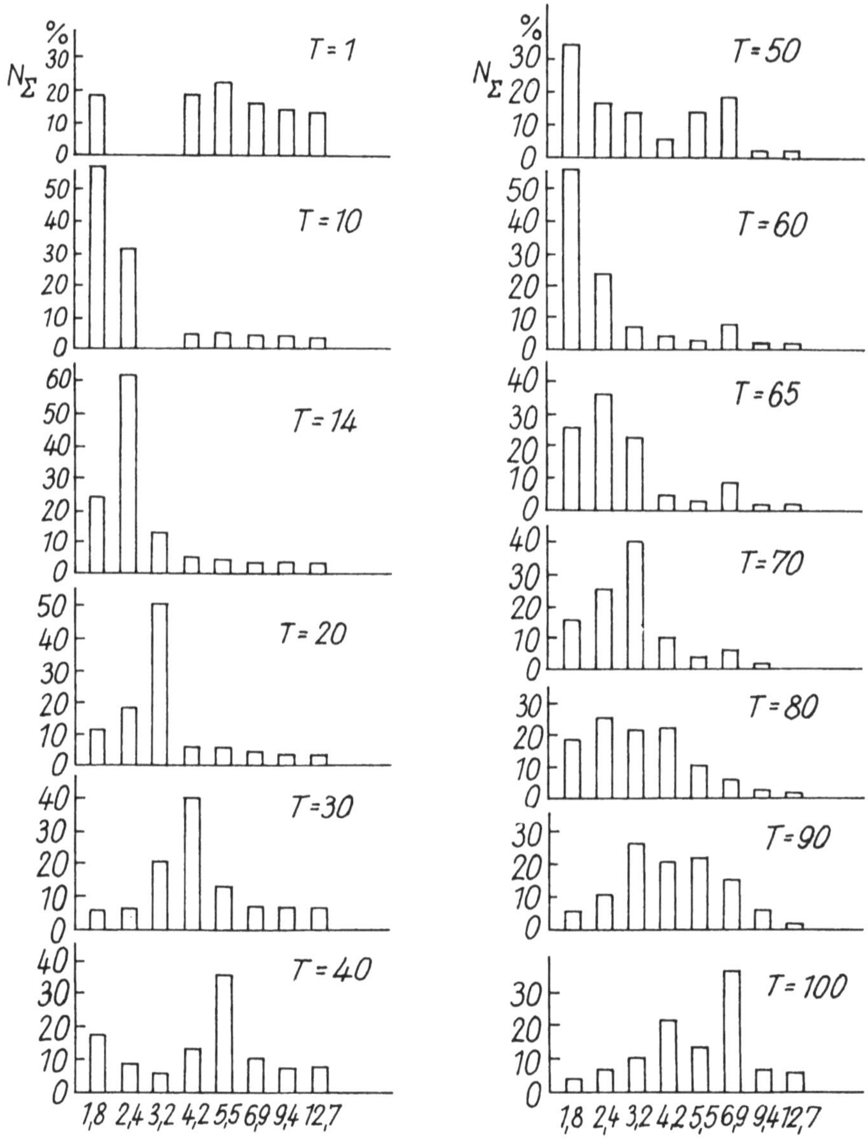

Fig. 23 The model crab population biomass:
T - model time, days; N_Σ - total number of crabs in population.

nearly the initial one (Gal'perin et al., 1981).

The increase of the crab population in this community reminds of a variety of population growth in which there is a time gap between the increased density of organisms because of the arrival of new specimens and the inhibiting action of this phenomenon on the number of specimens

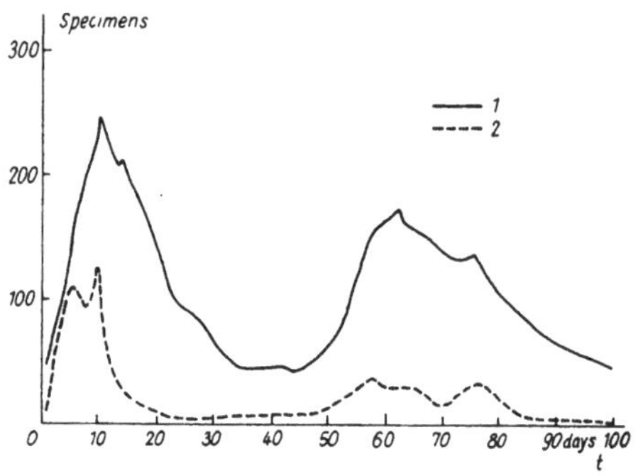

Fig. 24 Size groups of the model crab population.
(1) N - total and number of juveniles arriving daily (2), specimens /100 g of hydroid fouling.

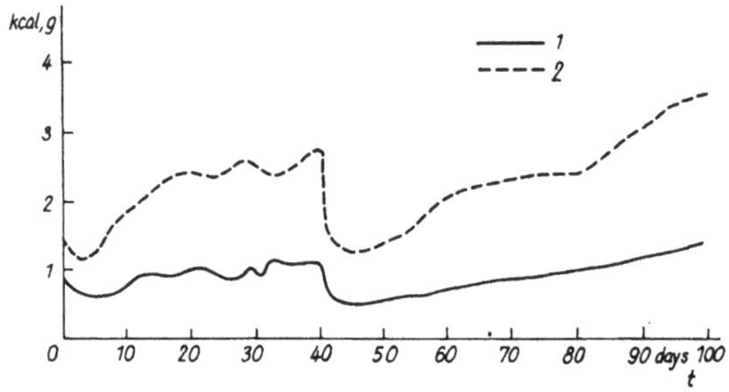

Fig. 25 Biomass of model crab population day weight (1), kcal (2) per 100 g of hydroid fouling.

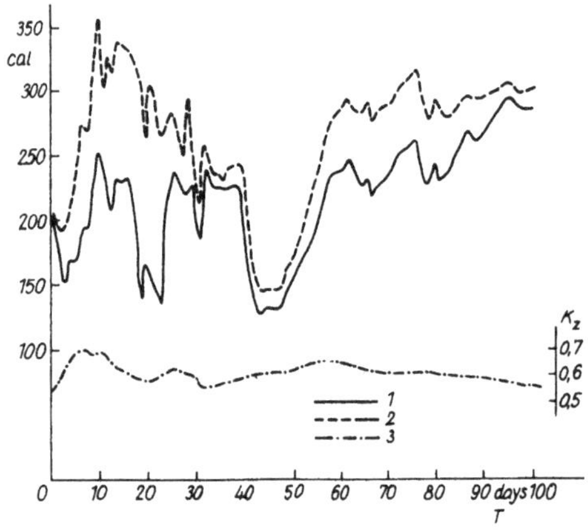

Fig. 26 Energy flow through a model crab population (1), assimilated food (2), factor K_2(3).

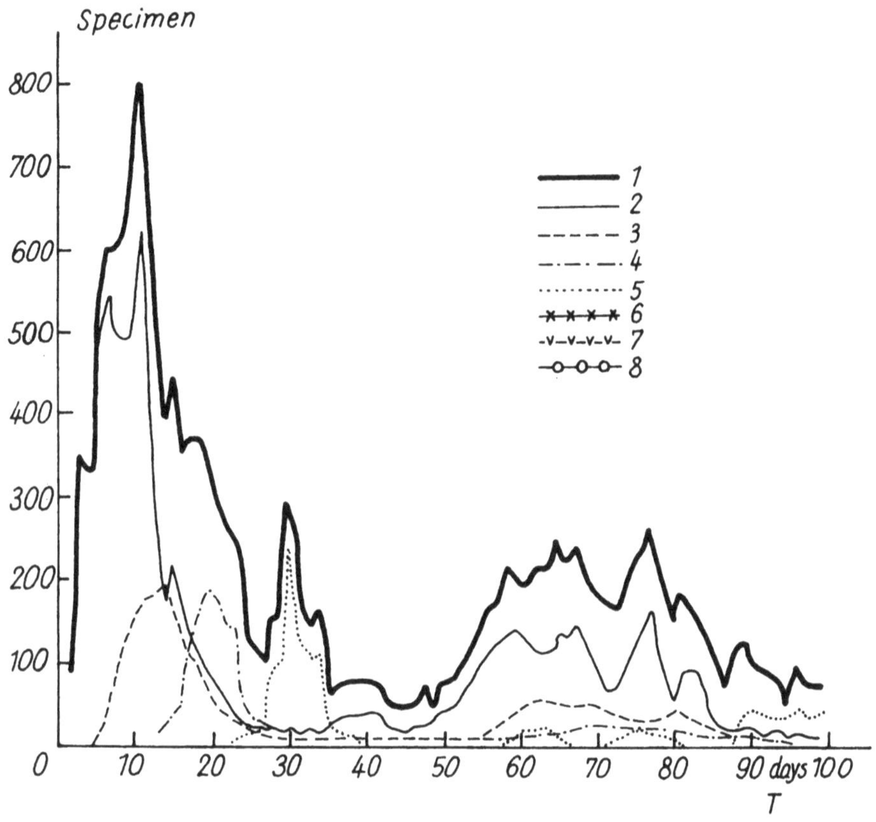

Fig. 27 Loss versus time in the total model crab population (1) and in size groups (specimens/day per one kg of fouling):
1 - 1.3 mm; 2 - 1.8 mm; 3 - 2.4 mm; 4 - 3.2 mm; 5 - 4.2 mm;
6 - 5.5 mm; 7 - 6.9 mm; 8 - 9.4 mm.

(Odum, 1971; Smith, 1974; Holling, 1978). Population variations with time are strongly dependent on the arrivals of juveniles, Fig. 24. The biomass of the crab population in weight units (grams of dry weight) and in energy units (calories) is represented in Fig. 25. In the first half of the model time the population biomass increased from 0.5 to 1 g and from 1.2 to 2.7 kcal and in the latter half it increased gradually to 1.5 g (3.5 kcal) on the 100th day.

The energy flow PE(T) through the population in one day varies widely with time, Fig. 26. During the first 50 days it increases to nearly 300 cal/day, falls to 140 cal, then rises again to 240 cal to fall to nearly 130 cal/day. In the second half, it first increases sharphly to nearly 240 cal/day by the 60th day and then, while varying slightly, gradually increases to 280 cal/day. The amount of energy assimilated by this population in one day $AC_{\Sigma}(T)$ is higher than PE(T) is all the time. At different times the difference varies as the population develops. They are

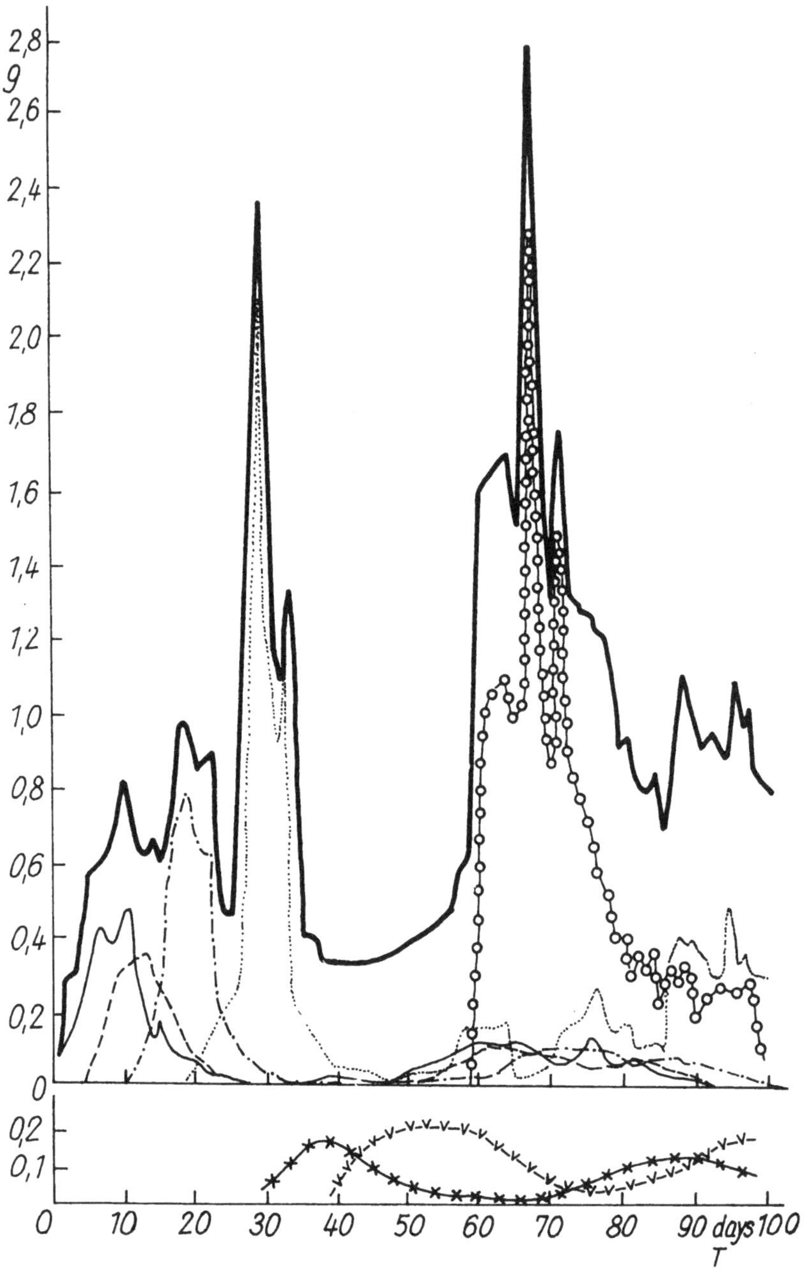

Fig. 28 Loss of model crab population biomass, g/day per one kg of hydroid fouling. Notation as in Fig. 27.

the largest when a great number of size 1.8 juveniles join the population. In particular, from the 10th to the 23rd day, when the juveniles grow fast (Gal'perin et al., 1981), AC_{Σ} (T) exceeds PE(T) by 100 to 160 cal; in other words, when large numbers of juveniles settle and grow,

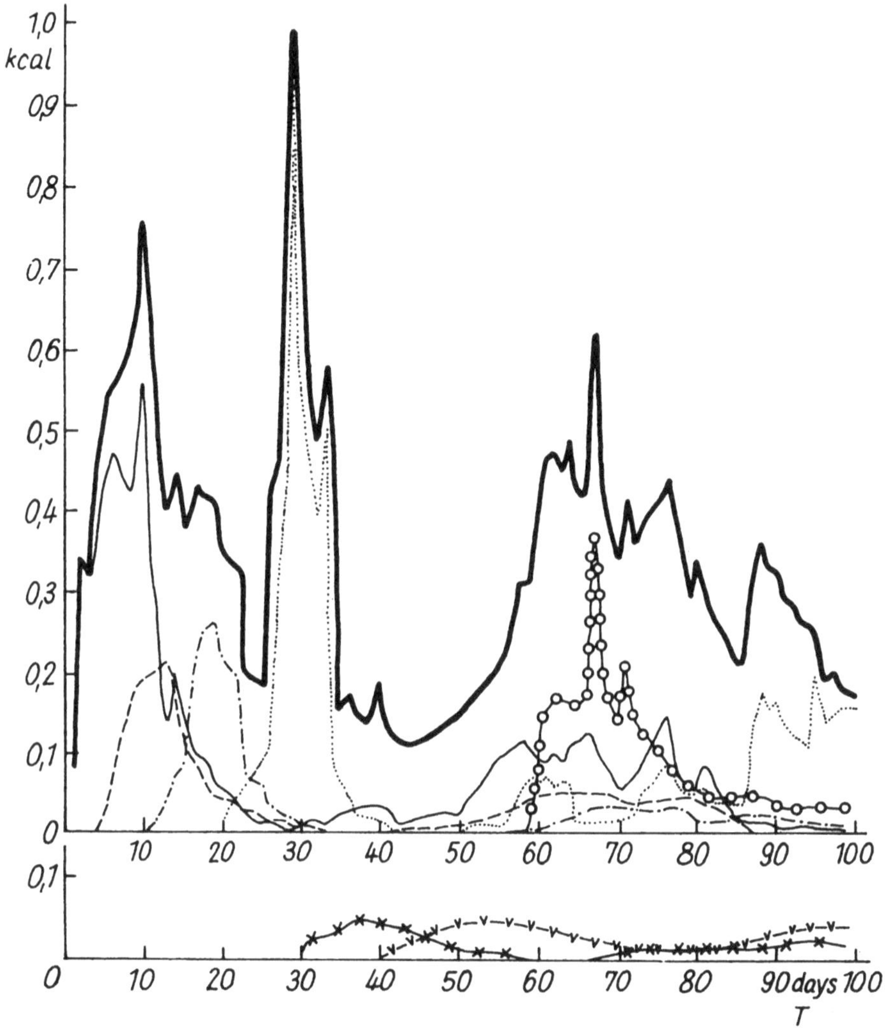

Fig. 29 Loss of energy in the model crab population (kcal/day per one kg of hydroid fouling). Notation is as in Fig. 27.

nearly one third of the daily food energy needs is satisfied by cannibalism. The daily values of K_2, a factor characterizing the efficiency of the crab population growth, oscillate near 60 per cent all the time and increase to 70 - 77 per cent during the juvenile grwoth and fall to 55 - 58 per cent as the number of large crabs grows in the population (Gal'perin et al., 1981; Chugay et al., 1981).

The number, weight, and energy characteristics of eliminated crabs are shown in Figures 27 - 29. The energy consumed by the activities of specimens which die later may be viewed as a loss. In this case, some of this energy, 8.5 kcal per 100 g of the fouling, must be subtracted from the total loss, which thus decreases to 11.0 kcal, or nearly half the energy flow through the model population in 100 days. The technique of computing the energy re-use in cannibalism neglected the energy that the population obtains by feeding on crabs larger than 3.2 mm, whereas large crabs, in fact, devour some specimens of their own size, in particular during molting, but this energy could not be measured physically (Chugay, Gal'perin, and Turpayeva, 1981; Turpayeva et al., 1981).

The model has thus shown the applicability of simulation to studies of populations of mobile animals in fouling communities. A satisfactory fit of the model and actual data made it possible to stage computer experiments and obtain results which shed light on the dynamics of the crab population in the community. Moreover, the model also determines the dynamics of weight and energy characteristics of the crab population and estimates the costs of self-regulation. Research and the quantitative estimation of elimination and migration processes make the analysis of the ecological impact the crabs have on the fouling community and their contribution to trophodynamic processes feasible. As this population is omnivorous, the numerical results on the energy flow can be extended to the interaction of species as a function of the substrate areas taken up by the population. The parametric independence of the modules and a good fit of the simulation results and control data explain why the model software was included in the simulation model of the community. The accuracy of numerical simulation results will be discussed in detail in Sect. 2.5.

2.5. Statistical Analysis of Empirical Relations and Results from Autonomous Models

As noted in Chapter 1, the adequacy analysis of all the models was performed as a quantitative and qualitative comparison of simulation results and field observations. The mathematical tools applied in formalization require numerous empirical relations that were embodied in the first and second hierarchy levels, Section 1.4. Therefore, the significance of simulating quantities at the higher levels of structural flowcharts depends on the accuracy of these relations. Detailed structuring of the system to be modelled and the specifics of available field data made it possible to use the method of least squares on most occasions (Brandt, 1970; Afifi and Eisen, 1979) in order to choose the relations and their parameters. The system of normal equations solved led to the desired coefficients of empirical relations in using this method in the form.

$$b_0 = \frac{\sum y \sum x^2 - \sum xy \sum x}{n \sum x^2 - (\sum x)^2}, \qquad (2.5.1.)$$

$$b_1 = \frac{n \sum xy - \ x \sum y}{n \sum x^2 - (\sum x)^2}. \qquad (2.5.2.)$$

Because this system of equations is applied in case of pairwise linear regression, the actual kind of the empirical dependence dictates the kind of linear transformations: for exponential power dependences, lg y and lg x, while for exact exponential dependences, ln y, x. The expressions for pairwise regression coefficients in these cases are lg b_0, b_1 and ln b_0, b_1, respectively. The significance boundary is established with the use of percentage distribution of Student points from the relation

$$\hat{t} = |b| / \bar{S}_b \geqslant t^T_{(n - 2;\ p)}, \qquad (2.5.3.)$$

where b is the selective estimate of the regression coefficient; $\bar{S}_b$ is the standard error; and t^T is a tabulated value. If the inequality (2.5.3.) is true, the coefficient is assumed to be significant according to the associated significance level.

In processing the empirical material obtained in unique experiments, the estimation of the entire regression dependence is very important. This is usually done by the Fisher criterion (Brandt, 1970; Afifi and Eisen, 1979). The general variance $\bar{S}^2$ and residual variance $\bar{S}^2_{res}$ are usually computed. The latter shows the difference between the value predicted by the regression equation and the experimental data. For the p-th significance level the following relation must hold

$$\hat{F} = \bar{S}^2 / \bar{S}^2_{RES} \geqslant F^T_{(n - 1,\ n - 2;\ p)}, \qquad (2.5.4.)$$

where F^T is the tabulated value of the Fisher criterion and $\hat{F}$ is the computed value. In research projects in simulation, where numerous empirical relations are applied special attention is paid more to the credibility of the dependence than to the accuracy of regression coefficients. This is due to the uniqueness of most field observations of natural processes, the amount of effort needed to make these observations repeatable, and the sensitivity of the observations to small perturbations. Power and exponential dependences are used in most cases, the former if the quantities are related indirectly. Exponential dependence is characteristic of cases when quantities are related directly. The choice of the dependence form is dictated in numerous ecological research projects, including this one, by the biological view of the processes. Combined with the specifics of field data for the system, this approach makes a detailed study of the free term of regression unnecessary because in most cases this is the initial number on experimental plates or the weight of the first polyp, larvae, etc.

The tables 17 - 19 summarize the statistical processing of field data applied to the hydroid, balanus, and infusorian models. The tables show the values of main coefficients, their significance, and regression.

For the hydroid population, the most important indicator is the ratio for computing the rate of the colony growth as a function of the plankton concentration. The significance of the remaining dependences other than the relation for the energy value of gonophores was five per cent because of the limited number of available measurements. The nature of the regression is specified by the coefficients with an acceptable accuracy. The empirical relations in the autonomous model of the balanus population describe the experimental results more accurately than the hydroid models. This can be seen in the row of the significance of regression equations. The significance is mostly checked by using the five per cent coefficient. For nearly all expressions this level is maintained, the gonophore energy value relations being an exception.

The significance of the equation for the growth of the infusorion colony may be investigated, because of the low number of measurements (three test points), only by the mean error. In statistical analysis, errors or residues are differences between the observed value of the quantity (test point) and the regression equation, which provides a theoretical description of the response function, with the main requirement that the mean error is approximately zero. The r.m.s. error in percentage of the test points may also be useful. The prediction or calculated value is regarded as satisfactory if this error does not exceed 15 per cent, which is the case of Table 20.

The data on weighted mean errors are especially valuable for the crab population because an "inverse" method was employed in the development of the model, Sect. 2.4. Recall that the empirical table of observations on the age structure and the biological ideas on the characteristic types of dynamics for various age groups during the first 12 days led to coefficients of the associated equations. Then the entire dynamics of the crab population was simulated over the whole model period. For this reason the data of Table 20 illustrate the significance of coefficients in the ratios to be used in computing the variations in the same age group, as well as the accuracy and significance of the computation.

Let us try now to estimate the impact made by changing the estimation coefficients in the empirical relations on the results to be obtained from the simulation of the population life. A change of the coefficient in the ratio for the hydroid rate of growth by as little as 0.04 results in the simulation results on the population biomass deviating by six per cent from the value obtained with the coefficient assumed in this research. Since the biomass deviation from the test points is 12 to 15 per cent, Sect. 2.1., this sensitivity seems good. A similar conclusion may be made for the coefficient in the ratio of the maximum biomass of the regenerative hydroid part. In this case, a change by the confidence interval, 0.09, results in a 13 to 18 per cent deviation and the scatter of test points around the mean is 15 to 20 per cent. The balanus population model was found to be highly sensitive to changes of the coefficients. The deviation in percentage in response to changes in the coefficients of the empirical rations is commensurable with the scatter of field data. For the nudibranch population changes of the parameters are significant only during the first eight to 12 days of the model time. Deflections of the coefficients of empirical ratios by the confidence interval change the results for this period by 20 to 27 per cent for the population fertility. Subsequently, this effect falls to 12 to 18 per cent. Changes of the coefficients for the dynamics of various age groups in the crab population increase the weighted mean error to a maximum of 40 per cent of the tabulated value and so the coefficients seem to have been chosen from field data satisfactorily. Increasing the mean weighted error in response to numerical changes of the coefficients proves that the model is sensitive and the results are not biologically impossible. Consequently, the sensitivity of the model for crabs seems to be admissible.

The statistical analysis of the significance of empirical relations applied to autonomous models, the analysis of the consistency of the simulation and field data for quantities obtained at every hierarchical level (Section 2.1. - 2.4.) and qualitative similarity of the behaviour of model and actual systems prove that the simulated behaviour is adequate and the numerical results are significant.

Table 17 Statistical estimation of coefficients of basic ratios for computing the hydroid energy balance

Ratio and its equation / Quantity	Growth rate (2.1.1.)	Maximal weight of generative part (2.1.3.)	Energy value, gonophores (2.1.13.)	Number of hydranths (2.1.14.) at temperature 20°C	Number of hydranths (2.1.14.) at temperature 24°C
$\gamma_1, \gamma_2, \gamma_3, \gamma_4$	0.5	1.1	1.28	0.85	1.21
$\bar{S}_{\gamma_1,\gamma_2,\gamma_3,\gamma_4}$	0.24	0.68	0.52	0.45	0.64
$\hat{t}_{\gamma_1,\gamma_2,\gamma_3,\gamma_4}$	2.083	1.620	2.449	1.879	1.891
$t^{T}_{(n-2;\ p)}$	2.074	1.372	1.886	1.859	1.859
n - 2; p	22; 2.5%	10; 10%	2; 10%	8; 5%	8; 5%
$\bar{S}^{2}$	0.108	0.040	0.171	0.101	0.144
$\bar{S}^{2}_{RES}$	0.037	0.013	0.019	0.028	0.041
$\hat{F}$	2.918	2.981	9.165	3.560	3.621
$F^{T}_{(n-1;n-2;p)}$	2.749	2.913	9.162	3.388	3.388
n-1; n-2; p	23;22;1%	11;10;5%	3;2;10%	9;8;5%	9;8;5%

Table 18 **Statistical estimation of coefficients of basic relations for computing the dynamics of the hydroid population**

Ratio / Quantity	Growth rate uncrowded settlement		Growth rate, crowded settlement		specific	Number		Critical
	juvenile	adult	juvenile	adult	mean, multi-layer settlement	before change of diet	after change of diet	Multilayer
Equation	(2.2.1.)	(2.2.1.)	(2.2.1.)	(2.2.1.)	(2.2.2.)	(2.2.7.)	(2.2.7.)	(2.2.11)
CW_B, γ_6, γ_7, γ_8	0.60	0.81	0.51	0.01	2.43	0.4	0.6	0.49
$\bar{S}_{CW_B}$, $\gamma_6 - \gamma_8$	0.193	0.021	2.112	0.003	1.56	0.199	0.262	0.278
$\hat{t}_{CW_B}$, 6 - 8	3.102	2.975	4.542	3.801	1.557	2.008	2.291	1.763
$t^T_{(n-2;\ p)}$	2.998	2.897	4.032	3.355	1.383	1.943	2.132	1.761
n - 2; p	7; 1 %	8; 1 %	5; 0.5%	8; 0.5%	9; 10%	6; 5%	4; 5%	14; 5%
$\bar{S}^2$	6.545	0.165	5.536	0.129	0.303	0.102	0.105	0.023
$\bar{S}^2_{RES}$	1.023	0.078	0.636	0.029	0.639	0.022	0.002	0.008
$\hat{F}$	6.393	4.721	8.702	4.362	4.075	4.649	6.382	2.873
$F^T_{(n-1;n-2;p)}$	4.899	4.357	6.977	4.357	3.347	4.207	6.256	2.462
n - 1; n - 2;	8; 7;	6; 5;	9; 8;	9; 8;	10; 8;	7; 6;	5; 4;	15; 14;

Table 19 Statistical estimation of coefficients of basic ratios for the computation of the nudibranch specimen

Ratio and its equation	Linear growth (2.3.1.)		Fertility (2.3.2.)	
	with salinity		with salinity	
Quantity	5 o/oo	8 o/oo	5 o/oo	8 o/oo
C_5; C_7	1.18	1.21	5.07	5.45
$\bar{s}_{C_5, C_7}$	0.384	0.403	2.035	1.960
$\hat{t}_{C_5, C_7}$	3.071	2.978	2.491	2.781
$t^T_{(n-2; p)}$	2.940	9.921	2.365	2.306
n; p	17; 0.5 %	18; 0.5 %	9; 2.5 %	10; 2.5 %
$\bar{s}^2$	0.376	0.536	5.109	5.995
$\bar{s}^2_{RES}$	0.149	0.217	1.364	1.752
$\hat{F}$	2.521	2.468	3.747	3.421
$F^T_{(n-1;n-2;p)}$	2.404	2.352	3.726	3.388
n - 1; n- 2; p	16;15;5%	17;16;5%	8;7; 5%	9;8; 5%

Table 20[*)] Mean computation errors for basic characteristics of the crab and infusorian populations

Number, size group		n	$\bar{e}_0 = \frac{\sum(y_T - y_7)/y_7 \cdot 100}{n}$ per cent
Crab population age and size groups	1.8 mm	16	2.67
	2.4 mm	16	2.09
	3.2 mm	14	7.78
	4.2 mm	15	12.53
	5.5 mm	12	12.24
	6.9 mm	13	9.07
	9.4 mm	12	5.11
	12.7 mm	13	5.03
Unidimensional infusorian colony		3	9.29

*) Computed from (Gal'perin et al., 1981).

Chapter 3. SIMULATION STUDY OF THE FOULING COMMUNITY BY THE PROPOSED APPLIED METHODS

3.1. Flowchart and Description of Interspecific Relations in the Community

As noted above, in the fouling community of the Sea of Azov, as in other bottom communities, topic as well as trophic relations have to play a major role. A qualitative flowchart of interspecific relations in the community is shown in Fig. 30 (Gal'perin, 1983a-e). The hydroid dominates the biocoenosis, the balanus being the runner up. The larvae of these species usually settle in late spring or early summer and develop in parallel. Under favourable conditions and with the plankton in good supply the hydroid covers the substrate in a thick layer as fast as within a month. Its branching stolons cover the balanus and thus cut their number. Soon the infusorian develops on the stolons and the inhibition of the balanus becomes stronger. The nudibranch joins the community in June and also stays on the hydroid stolons and feeds on the hydroid and infusorian hydranths. As long as it is small and dominated by juvenile specimens, the nudibranch population feeds chiefly on the infusorian; if it is large, not only does it consume all infusoria but also inhibits the hydroid to such an extent that parts of the colony are deprived of the hydranths and so separated from the food. As a result, the balanus feeding conditions are improved, which also occurs at lower salinity because in this case the hydroid growth rate diminishes dramatically. The reduction of the substrate area for the development of the infusorian and a new nudibranch generation when parts of the hydroid colony break away culminates in an intraspecific competition in the nudibranch population. In July, crab larvae settle in large numbers, juvenile crabs feed on the infusorian and, when they grow to medium size, also on the nudibranch. Large crabs feed on the nudibranch eggs and, to some extent, on their own juveniles. The crab has been found to migrate both inside and from the community with water currents.

The community produces at a high rate in late spring and early summer, and this 100-day period is especially enlightening. As reported in Chapter 1, field data were obtained from experimental 1 m^2 plates where either isolated populations (others being scraped off) or groups of cohabitating species developed. These observations made it possible to use, in addition to quantitative data, qualitative (heuristic) knowledge of the mechanisms of population interaction, formation, and development.

A detailed description of the qualitative interspecific relations in the fouling community was also reported elsewhere (Gal'perin, 1983 a-e). In addition to the basic classification into trophic and topic, topic relations were distinguished both by the time of their functioning and

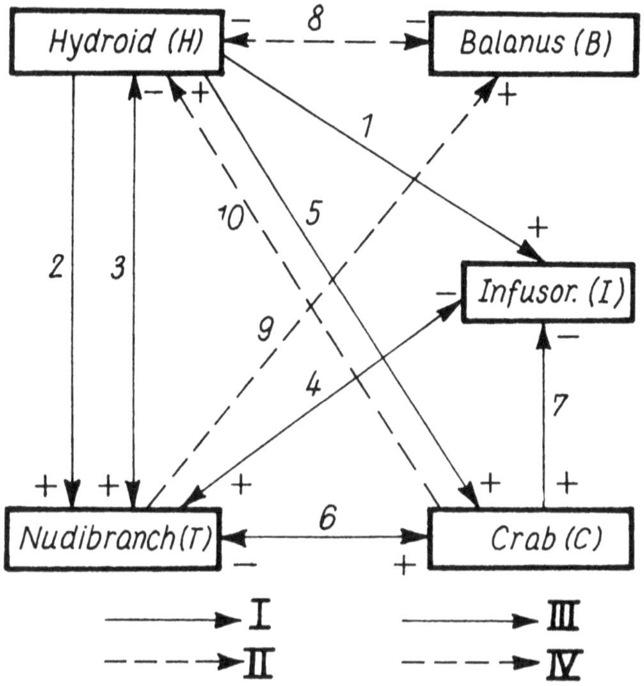

Fig. 30 Qualitative flowchart of symphysiological relations in the fouling community.

the mechanisms through which they influenced the development and vital functions of the populations. The processes controlled by natural forces were the growth rate of hydroid and balanus; the dynamics of the nudibranch population; the same of the crab and infusorian; energy flows through the populations; and the dynamics of relations between elements of the energy balances. Energy indicators, as noted in Chapter 1, were used for the quantitative description and comparison of these various processes and the rates of interspecific relations. The rates of interspecific relations were quantitatively estimated as the values of loss or, alternatively, as benefit from the interspecific relation. With this interpretation, these rates may be estimated in terms of weights (kg, g, mg) and the numbers of specimens per unit weight or area. What is important is that Fig. 30 shows only the relations that are most significant in the model period and have a remarkable impact on the formation and functioning of the community. The entire system of interspecific relations may be analyzed as a totality of interacting groups of relations which may be either topic or trophic as is the case of the nudibranch's group of relations (Nos 2-4). The interspecific relations

Table 21 Interspecific relations of populations in the simulated community

Ordinal number	Interacting populations	Model time of action (days)	Classification	Process
1	Hydroid-infusorian	25 - 100	direct topic	Provision of substrate for most of the infusorian population.
2	Hydroid-nudibranch	36 - 100	direct topic	Hydroid provides substrate for the nudibranch population to develop and stay.
3	Nudibranch-hydroid	36 - 100	direct trophic	Food supply for nudibranch population; consumption of hydranths and ensuring restructuring of elements in hydroid population energy budget. Inhibition of hydroid, separation of damaged parts of colony.
4	Nudibranch-infusorian	36 - 100	direct trophic	Partial feeding of juvenile nudibranches. Reduction of the number of infusorian heads.
5	Hydroid-crab	36 - 100	direct trophic	Provision of substrate for the entire crab population
6	Crab-nudibranch	36 - 100	direct trophic	Feeding crab population partially (large crabs). Damage to nudibranch eggs.
7	Crab-infusorian	36 - 100	direct trophic	Feeding some of the crab juveniles.
8	Hydroid-balanus	5 - 45	indirect topic	Competition for substrate. Reduced growth of balanus. Decreased growth of hydroid in response to change of salinity and formation of multilayer balanus settlement.
9	Nudibranch	36 - 45	indirect trophic	Reduction of pressure on balanus with consumption of hydroid.
10	Crab-hydroid	36 - 100	indirect trophic	Reduction of pressure on hydroid through inhibition of nudibranch population.

of populations are summarized in Table 21 and the functioning of the entire community is described in Section 1.3. and in this Chapter.

3.2. The Simulation Model of the Fouling Community. A Mathematical Description

General. The method to develop and implement the simulation model of this book starts with the investigation of autonomous models of populations and proceeds by introducing interspecific relations to simulate the entire process of community functioning. As noted in Chapter 1, a set of models of interspecific relations is used with a view on modelling the interaction of populations. Unlike the widespread methods in which simulation models are formalized by introducing the coefficients of interactions in some forms, in this case the interaction is represented as the potential of the population. The processes of interspecific relations are quantified in the light of ecological assumptions and knowledge. As a result, the formalization tools are systems of logically complex relations and algorithms for the computation of quantities. In most cases the following algorithm, which represents a trophic relation in the community, works:

1. Computation of the potential needs of the predator population for certain conditions and its status in an autonomous mathematical model.
2. Computation of the possible needs and potential of the prey population.
3. Checking the conditions for the realization of the interspecific relation and computing the relation characteristic.
4. Simulation of interaction and recomputation of the characteristics of the population and relation.
5. Recomputation of indirect relations for the new population state.
6. Transfer to new pairs of interacting populations. For topic relations, a unified algorithm of realizing the interspecific relations may be reduced to the following steps:

(1) On the knowledge of ecological values, compute areas where the dependent species develop.

(2) Computing, in the autonomous mathematical model, the potential needs of the dependent population per unit area of the conditioning species.

(3) Simulating the functioning of the dependent population on the entire substrate area available.

(4) Recomputing the appropriate quantities if the topically interacting populations fall prey to a predatory species.

(5) Computing the quantitative characteristics of topically interacting populations from actual population states.

Consequently, the system of mathematical models of species interaction in a fouling community is made up of three types: (1) autonomous models, which compute the potential characteristics of interacting populations; (2) models of exogenous impacts of interspecific relations; and (3) models for recomputing the population characteristics if the interspecific relation influences the physiological processes in comparison with the autonomous model (endogenous impact). In the mathematical formalization of the system the superscript "*" will denote quantities whose method of computation or associated equation is described in Chapter 2. Equations used in the mathematical description of the system of models include components from the mathematical description of autonomous models, Chapter 2. The substript "S" will denote quantities obtained in the models for recomputing the population state and the superscript will denote the ordinal number of the relation in the flowchart of Fig. 30 and so in its model which simulates the impact on dependent populations. In this way, the logical sequence of computing the quantities is preserved, otherwise equations of models would be needed for every individual relation, which would be less legible.

The hydroid. The assimilated food ($\overset{S}{A}_H$) of the population is obtained from the system of input-output energy equations of Section 1.4., where in this case the addends are the growth energy ($\overset{S}{P}_H$), the metabolic energy ($\overset{S}{R}_H$), and the regeneration energy ($\overset{S}{G}_H$). Allowance should be made for the consumption ($E^{(3)}_{AH}$) by the predatory nudibranch and for the increase, wherever possible, of assimilated food ($\Delta U^{(3)}_H$) of the population for compensating the regeneration energy of consumed hydranths. This is feasible because the hydroid polyp colony is capable of partially or completely restructuring the intraspecific nutritional regulation and increasing the number of feeding hydranths.

$$\overset{S}{A}_H = \overset{S}{P}_H + \overset{S}{R}_H + \overset{S}{G}_H + \Delta U^{(3)}_H - E^{(3)}_{AH} \ . \qquad (3.2.1.)$$

The growth energy is

$$\overset{S}{P}_H = P^*_H \cdot \overset{S}{E}_{SH}(S) - \Delta P^{(3)}_{\Sigma} \ , \qquad (3.2.2.)$$

where $\overset{S}{E}_{SH}(S)$ is the coefficient of growth slow-down in response to salinity decrease (computed as percentage of the growth rate from field data) and $\Delta P^{(3)}_{\Sigma}$ is loss of the growth energy as a result of the nudibranch inhibiting the hydroid

$$\Delta P^{(3)}_{\Sigma} = \begin{cases} P^*_H & P^*_H \leqslant A^{(3)}_{REG} - A^*_{POT} - G^*_H \\ A^{(3)}_{REG} - A^*_{POT} - G^*_H & P^*_H > A^{(3)}_{REG} - A^*_{POT} - G^*_H \\ 0 & 0 \geqslant A^*_{POT} - A^{(3)}_{REG} - G^*_H \ , \end{cases} \qquad (3.2.3.)$$

where A^{*}_{POT} is the potentially assimilated food of the population for given environmental conditions, and $A^{(3)}_{REG}$ is the energy the population spends for the regeneration of colony parts where the hydranths have been consumed by the predator,

$$A^{(3)}_{REG} = \sum_{L=1}^{L_m} N_V^{(3)} (T - L) \cdot \varphi G(L) , \qquad (3.2.4.)$$

where φG is the regeneration energy of one hydranth ($L_m = 3$); T is the model time; and $N_V^{(3)}$ is the number of consumed and unregenerated hydranths,

$$N_V^{(3)} = \hat{A}^S_T / K^*_H . \qquad (3.2.5.)$$

In (3.2.5.) $\hat{A}^S_T$ is the fraction of assimilated hydroid food of adult nudibranches and K^*_H is the energy value of one hydranth.

The metabolism energy is computed as in autonomous models, Chapter 2, and the regeneration energy may be computed from (3.2.6.) and (3.2.7.)

$$G^S_H = G^*_H - \Delta G^{(3)}_H , \qquad (3.2.6.)$$

where $\Delta G^{(3)}_H$ is the loss of regeneration corresponding to the restructuring of the population energy budget under the impact of the nudibranch

$$\Delta G^{(3)}_H = \begin{cases} G^*_H & A^{(3)}_{REG} - A^*_{POT} \geqslant G_H \\ A^{(3)}_{REG} - A^*_{POT} & A^{(3)}_{REG} - A^*_{POT} < G_H \\ 0 & A^{(3)}_{REG} - A^*_{POT} \leqslant 0 . \end{cases} \qquad (3.2.7.)$$

The population energy budget has to be restructured because, under certain conditions, the energy the hydroid needs for the regeneration of lost parts of the colony cannot be compensated for by reducing the intraspecific feeding regulation. Indeed, when the feeding regulation indicator δ reaches its maximum, the hydroid population cuts the energy allocation for regeneration by the necessary amount. The fraction of regeneration in the energy flow falls and so the energy budget may be said to be restructuring. If this restructuring proves insufficient, the growth rate decreases. Shortage of the metabolic energy results in the separation of a part of the colony. These processes are formalized by a system of equations (3.2.8.) - (3.2.12.) and the relations (3.2.3.) till (3.2.7.).

$$E_{AH}^{(3)} = \begin{cases} 0 & \delta < 1 \\ N_V^{(3)} \cdot r^{/*}(\beta) & \delta = 1 , \end{cases} \qquad (3.2.8.)$$

$$\frac{d\tilde{N}_E^S}{dT} = \frac{dN_E^*}{dT} - \left[E_{AH}^{(3)} - \Delta U_H^{(3)} \right] / \; r^{/*}(\beta) , \qquad (3.2.9.)$$

$$\frac{d\tilde{N}_H^S}{dT} = \frac{dN_H^*}{dT} - N_V^{(3)} + \Delta U_H^{(3)} / \, r^{/*}(\beta) , \qquad (3.2.10.)$$

$$\delta = \tilde{N}_E^S / \tilde{N}_H^S , \qquad (3.2.11.)$$

$$\Delta U_H^{(3)} = \begin{cases} A_{REG}^{(3)} & A_{REG}^{(3)} \leqslant A_{POT}^* \\ A_{POT}^* & A_{REG}^{(3)} > A_{POT}^* . \end{cases} \qquad (3.2.12.)$$

In the system (3.2.8.) - (3.2.12.) r'^* is the individual ration of one hydranth that depends on the plankton food β per cubic meter of water; N_E^* and N_H^* are the potential variation of the number of feeding and live hydranths, respectively. The equation for the population weight dynamics in energy units is

$$\frac{dW_H^S}{dT} = P_H^S - V_1^{(3)} - V_0^* - V_2^{(3)} , \qquad (3.2.13.)$$

where V_0^* is the energy value of reduced and separated parts of the colony after the regeneration, and the energy value of the consumed hydranths is

$$V_1^{(3)} = N_V^{(3)} \cdot K_H^* . \qquad (3.2.14.)$$

The energy value of the separated part is expressed in case of energy shortage for energy metabolism due to shortage of energy for compensating the regeneration as

$$V_2^{(3)} = \begin{cases} 0 & A_H^S \geqslant R_H^S \\ \Psi_H^* \cdot \left[R_H^S - A_H^S \right] / \left[TR^* \cdot N_0^* \right] & A_H^S < R_H^S , \end{cases} \qquad (3.2.15.)$$

where TR^* is the individual ration of one hydranth: Ψ_H^* is the energy value of one kilogram of the colony; and N_0^* is the number of hydranth in one kilogram of the regenerative part of the colony, Section 2.1.

When the balanus makes a multi-layer settlement, the hydroid is inhibited. The balanus reduces the hydroid growth rate of the colony from time t when the balanus population is critical (for a free development of the hydroid), which is easily computed algorithmically from the data of field experimentation. Since, in the field experiment, the balanus-inhibited hydroid did not grow to the weight at which intraspecific feeding regulation began, it seems plausible to assume that the population grows exponentially as in the first days of the juvenile development. The growth rate of such a population is

$$\bar{C}_W = \begin{cases} C_W^* & T < \bar{t} \\ C_W^* - \bar{C}_d / \left[T - \bar{t}\right] & T \geqslant \bar{t} \end{cases}, \qquad (3.2.16.)$$

where C_W^* is the growth rate of the hydroid in free conditions and $\bar{C}_d$ is the mean decrease of C_W^* computed from empirical data. The necessary values are easily obtained from (3.2.16.) with (3.2.2.), (3.2.3.), and (3.2.13.).

<u>The balanus.</u> The development of the Balanus improvisus population, which settles in one stratum on an artifical substrate, depends on the food concentration β of zooplankton for the hydroid and α of phytoplankton for the balanus itself. In the model of the hydroid - balanus relation (No. 8) this interaction is indirectly expressed as the relative indicator $m = \alpha / \beta$ (per m^3 of water). In a fairly general form the assimilated food of the balanus population is

$$A_B^S = \sum_{j=1}^{n} (B_j^* - EB_j^{(8)}), \qquad (3.2.17.)$$

where B_j^* is the j-th element of the potential energy balance (computed by the procedure of Section 2.2.); $EB_j^{(8)}$ is the energy value of the associated element of the population or loss of the specimen's energy budget because of the hydroid - balanus relation (No. 8).

$$EB_j^{(8)} = f_j^* (W_{BP}^{(8)}), \qquad (3.2.18.)$$

where f_j^* is the transfer function of the weight $W_{BP}^{(8)}$ of the balanus population or specimen by which it falls below its potential to the j-th element of the population energy budget

$$W_{BP}^{(8)} = W_{BP}^{(8)} (T - 1) \cdot D_u^{(8)}, \qquad (3.2.19.)$$

$$D_u^{(8)} = \left[C_1 \cdot m^{-C_2} \right] \cdot (W_H^S)^{-\left[C_3 \cdot m^{-C_4}\right]}. \qquad (3.2.20.)$$

In (3.2.19.) and (3.2.20.) $D_u^{(8)}$ denotes the unity pressure of the hydroid population per unit weight of the balanus population and $C_1 - C_4$ are parameters chosen for the relation by the computer using regression analysis. The population weight can be easily expressed by the weight of one specimen (ω_B^S) and the population size, in this case constant. With j = 1 being the growth indicator and Ψ_B^* the energy value of a weight unit,

$$\frac{d\omega_B^S}{dT} = (B_1 - EB_1^{(8)})/\Psi_B^*. \qquad (3.2.21.)$$

In case of a multi-layer population assumptions must be made for the age structure of the population and the intraspecific competition of the barnacles for the substrate. All basic characteristics of the population are expressed through the population size computed from the equations (3.2.22.) and (3.2.23.)

$$N_B^S = \sum_{\tau=1}^{T} \left[NS_\tau^*(T) - \widetilde{NG}_\tau^* (t_\tau) - NK_\tau^S \right], \qquad (3.2.22.)$$

$$NK_\tau^S = \begin{cases} 0 & N_B^S \leqslant NC^* \\ N_B^S - NC^* & N_B^S > NC^*. \end{cases} \qquad (3.2.23.)$$

In this system of equations, NS^* is the number of larvae settled on the Γ-th day; τ is the ordinal number of the age group; $\widetilde{NG}_\tau^*$ is the number of barnacles of the τ-th age group dying of natural causes at the age of t_τ; NK_τ^S is the number of barnacles dying because of the intraspecific competition; NC^* is the maximally possible (critical) number of barnacles in the population on the substrate area investigated. The values of NS_τ^*, $\widetilde{NG}_\tau^*$, and NC^* were computed algorithmically, the values of the parameters were determined from field data by an auxiliary module of the simulation program.

The nudibranch. In computing the elements of the energy balance for the nudibranch population, assumptions should be made for the consumption of some adults by the crab (NG_1^S), damage to egg deposits done by the crab (NG_2^S), death of juveniles in the intraspecific competition for the infusorian food (NG_3^S), and adults when the hydroid is in short supply (NG_4^S). Since the j-th element of the nudibranch energy budget can be expressed as its potential value BT_j^*,

$$BT_j^S = BT_j^* - \sum_{k=1}^{4} \sum_{\tau=1}^{t_k} NG_k^S \cdot b_{kj}^* (t_\tau) , \qquad (3.2.24.)$$

where b_{kj}^* is the j-th element of the energy budget for a nudibranch specimen (or the energy value of eggs) which died of the k-th cause at the age of t_τ. In order to determine t_τ, the smallest specimens were assumed to die first. The matrix of dying nudibranches was computed algorithmically in compliance with the above assumption and with

$$\Delta AC = \sum_{\tau = 1}^{t_k} NG^*_\tau \cdot \omega k^*_\tau , \qquad (3.2.25.)$$

where ΔAC is the part of the nudibranches assimilated by the crab or the total assimilated food of nudibranches who died in intraspecific competition; NG^*_τ is the number of dead nudibranches in the τ-th generation; ωk^*_τ is their energy value (if they represent the food) or assimilated food (if they die in intraspecific competition). The assimilated food of the entire population is fairly easily obtained

$$\overset{S}{A_T} = \sum_{j = 1}^{m} \overset{S}{BT_j} , \qquad (3.2.26.)$$

as other population characteristics, too.

<u>The infusorian.</u> The number of heads (N^S_Σ) of the dependent infusorian population, which consists of protozoa of the same age, is determined by their number per square centimeter of the substrate (hydroid stolons) (N^S_S), the number ($NV_S^{(4, 7)}$) of infusorian heads consumed by the nudibranch and crab with energy needs $\bar{A}_T^{(4)}$ and $\bar{A}_C^{(7)}$, respectively, and the number (N^S_{RES}) of restoring infusorian heads in areas damaged by the predator. The system of equations for the number of specimens in the population is

$$\frac{dN^S_S}{dT} = n^*_p \, (1 - N^S_S / N^*_{max}) \cdot N^S_S , \qquad (3.2.27.)$$

$$N^{(1)}_S = P^{(1)} \, (\Delta \overline{W}^{(S)}_H) \cdot N^S_S , \qquad (3.2.28.)$$

$$N^S_{RES} = \sum_{L_1 = t_1}^{T} \left[N^S_S \cdot P^{(4)} (\bar{A}^{(4)}_T (T - L_1)/(\Psi_S^* \cdot N^*_{max})) \right] +$$

$$+ \sum_{L_2 = t_2}^{T} \left[N^S_S \cdot P^{(7)} (\bar{A}^{(7)}_C (T - L_2)/(\Psi_S^* \cdot N^*_{max})) \right] , \qquad (3.2.29.)$$

$$NV_S^{(4, 7)} = (\bar{A}^{(4)}_T + \bar{A}^{(7)}_c)/\Psi_S^* , \qquad (3.2.30.)$$

$$\frac{dN^S_\Sigma}{dT} = N^{(1)}_S + N^S_{RES} - NV_S^{(4, 7)} , \qquad (3.2.31.)$$

where $P^{(n)}$ is the function for the transfer from energy and weight characteristics of interacting populations to the substrate (hydroid) areas where the infusorian can develop or restore; n_p^* is the coefficient of population growth; Ψ_S^* is the energy value of the infusorian head; N_{max}^* is the maximal number of infusoria per square centimeter of the substrate; t_1 and t_2 are times of death of associated colony areas which are consumed by the nudibranch and crab. The values of $\bar{A}_T^{(4)}$ and $\bar{A}_C^{(7)}$ were computed from the numbers of juveniles in the nudibranch population (of size less than 4.5 - 5 mm) and the crab population (size below 4.2 mm). The model also recognizes the alteration of periods when the infusorian is consumed by the nudibranch and periods of restoration at a high rate. This alteration is explained by the fact that the nudibranch leaves the damaged parts of the colony and the restoration takes two to four days after the predator has left. By the time the restoration is nearly complete, the nudibranch returns and the nudibranch - infusorian relation (No. 4) resumes.

The crab. This widely spread species in the fouling of the Sea of Azov is omnivorous and mobile. For this reason in most relations the crab is trophically independent and plays a conditioning role. The development of this population is contingent only on the weight ($\tilde{W}_H^S$) of the hydroid fouling, where the crab abides. The basic processes in the crab population and their model were discussed in Section 2.4. On the knowledge of the values of (BC_j^*) for the crab population per 100 g of the hydroid fouling obtained in the autonomous model, and bearing in mind that there is a nearly uniformly developed and thin substrate on this area, the characteristics ($\overline{BC}_j^S$) of the crab population can be obtained in a general form for the community model from the relation

$$\overline{BC}_j^S = BC_j^* \cdot \tilde{W}_H^S \quad . \qquad (3.2.32.)$$

Quantitative estimates of species interaction forces. The species interaction forces may be evaluated in absolute quantities (kg, kcal, specimens) or relative dimensionless quantities. For topic relations this force in absolute units depends on the substrate area of weight. More specifically,

$$Z_k^{(n)} = \tilde{\varphi}_k^{(n)} (P_Z) \ , \qquad (3.2.33.)$$

where P_Z is the substrate area and $\tilde{\varphi}_k^{(n)}$ is the function for recomputing the area into weight and energy characteristics. Then, in dimensionless units, the rate of the interspecific topic relation is determined by the formula

$$\xi_k^{(n)} = 1 - Z_k^{(n)} (T - 1) \,/\, Z_k^{(n)} (T) \ . \qquad (3.2.34.)$$

The trophic relations are estimated in absolute units as

$$ZF_m^{(n)} = \sum_{i=1}^{I} \Delta LS_{mi}^{(n)} , \qquad (3.2.35.)$$

where m is the indicator of the characteristic of a dependent population; i is the subscript of the exogenous process; and $\Delta LS_{mi}^{(n)}$ is the variation of the m-th characteristic (LS_m) of the dependent population as a result of the n-th relation for the i-th component. The relative rate of the trophic relation is

$$\xi F_m^{(n)} = ZF_m^{(n)} / LS_m \quad . \qquad (3.2.36.)$$

The system of basic equations (3.2.1.) - (3.2.36.) is embodied in the simulation model, logical-heuristic algorithms in particular, and in algorithms for the simulation program of the fouling community model in general.

3.3. Computer Simulation of the Fouling Community

Basic Algorithms

The simulation program of the fouling community is controlled by the head module which looks after the man - computer interaction; calls the modules of autonomous models; handles the data in external media; adjusts the simulation program; generates the sequence of calling the modules for computing the states of relations and populations following their interaction; prints-out controllable variables. This algorithm is represented by the blocks 1 - 8 and 11 - 15 in Fig. 31, which shows the general simulation algorithm for a final version of the simulation program in MVT 6.1 OS.

As noted in the Sections 1.2. and 1.4., logical-heuristic algorithms have to play an important role in simulation models. Such models are most successful when their structural flowchart is unified (Buslenko, 1977; Vorovich et al., 1981; Gal'perin and Keondchyan, 1981; Gal'perin and Turpayeva, 1982) and so the modules may be interfaced and the transfer from one level in the system hierarchy to another is algorithmical. The algorithms for the simulation of the fouling community are described in the Chapters 2 and 3. The figures 31 and 32 provide a deeper insight into two most important logical-heuristic algorithms, namely, the recomputation of the state of the hydroid population and of the state of the trophic nudibranch - hydroid state. For the steps of these

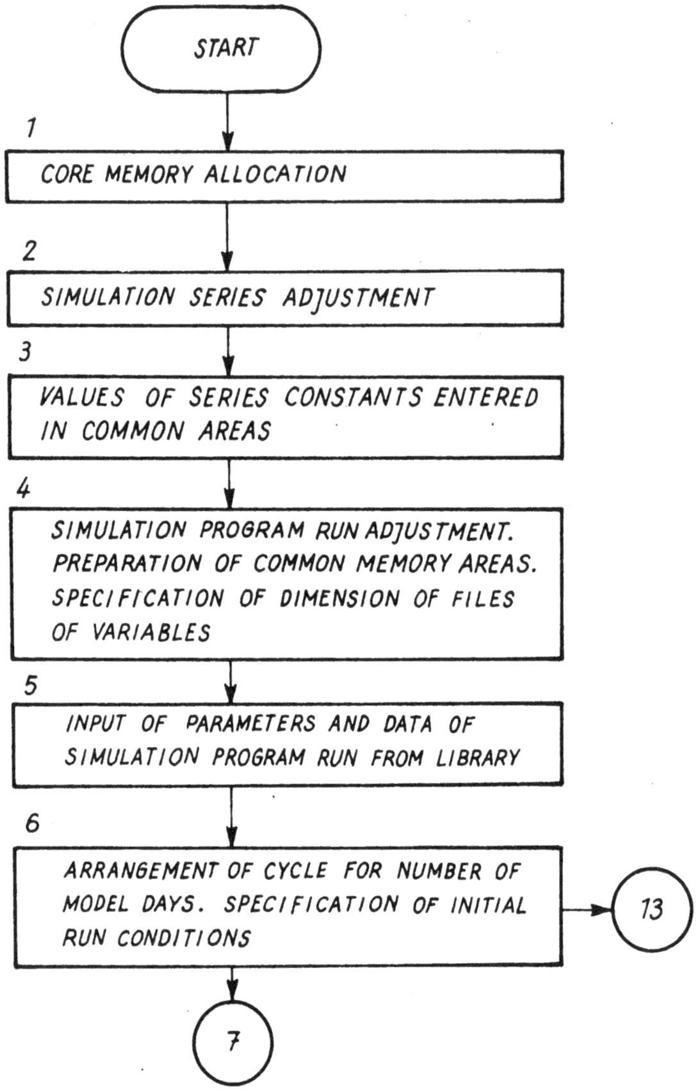

Fig. 31 Flowchart of the simulation program algorithm for a fouling community
0 - START; 1 - CORE MEMORY ALLOCATION; 2 - SIMULATION SERIES ADJUSTMENT; 3 - VALUE OF SERIES CONSTANTS ENTERED IN COMMON AREAS; 4 - SIMULATION PROGRAM RUN ADJUSTMENT. PREPARATION OF COMMON MEMORY AREAS SPECIFICATION OF DIMENSION OF FILES OF VARIABLES; 5 - INPUT OF PARAMETERS AND DATA OF SIMULATION PROGRAM RUN FROM LIBRARY; 6 - ARRANGEMENT OF CYCLE FOR NUMBER OF MODEL DAYS. SPECIFICATION OF INITIAL RUN CONDITIONS; 7 - CALL

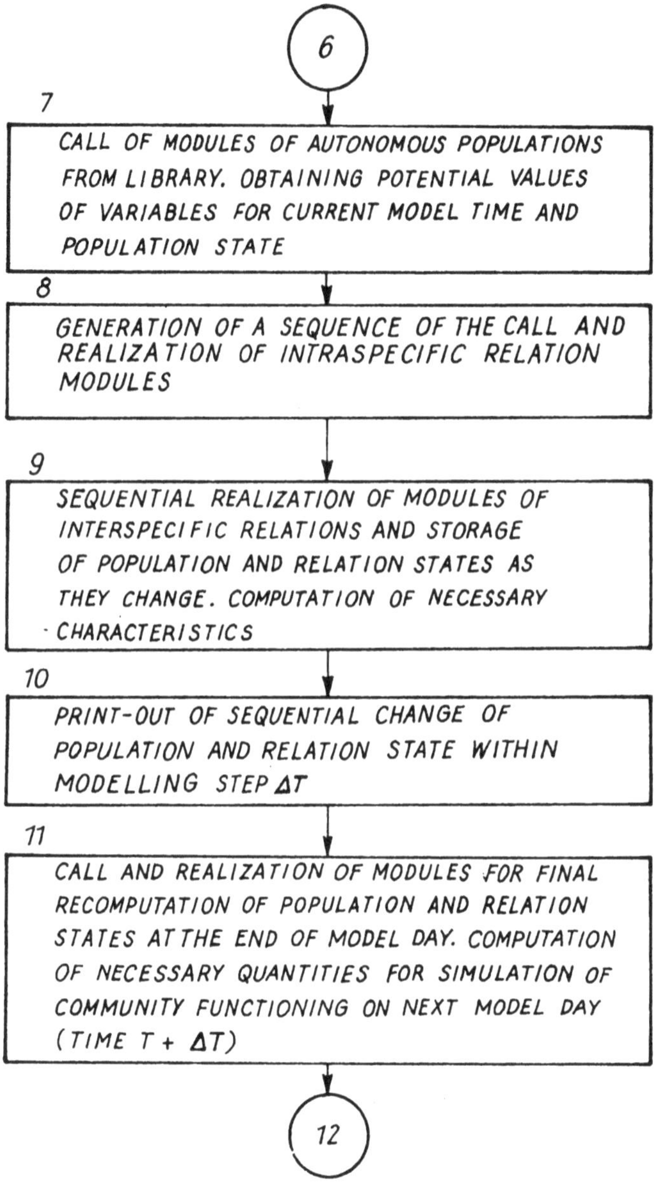

Fig. 31 (Continued)
OF MODULES OF AUTONOMOUS POPULATIONS FROM LIBRARY. OBTAINING POTENTIAL VALUES OF VARIABLES FOR CURRENT MODEL TIME AND POPULATION STATE; 8 - GENERATION OF SEQUENCE OF CALL AND REALIZATION OF INTRASPECIFIC RELATION MODULES; 9 - SEQUENTIAL REALIZATION OF MODULES OF INTERSPECIFIC RELATIONS AND STORAGE OF POPULATION AND RELATION STATES AS CHANGE. COMPUTATION OF NECESSARY CHARACTERISTICS; 10 - PRINT-OUT SEQUENTIAL CHANGE OF POPULATION AND RELATION STATE WITHIN MODELLING STEP ΔT; 11 - CALL AND REALIZATION

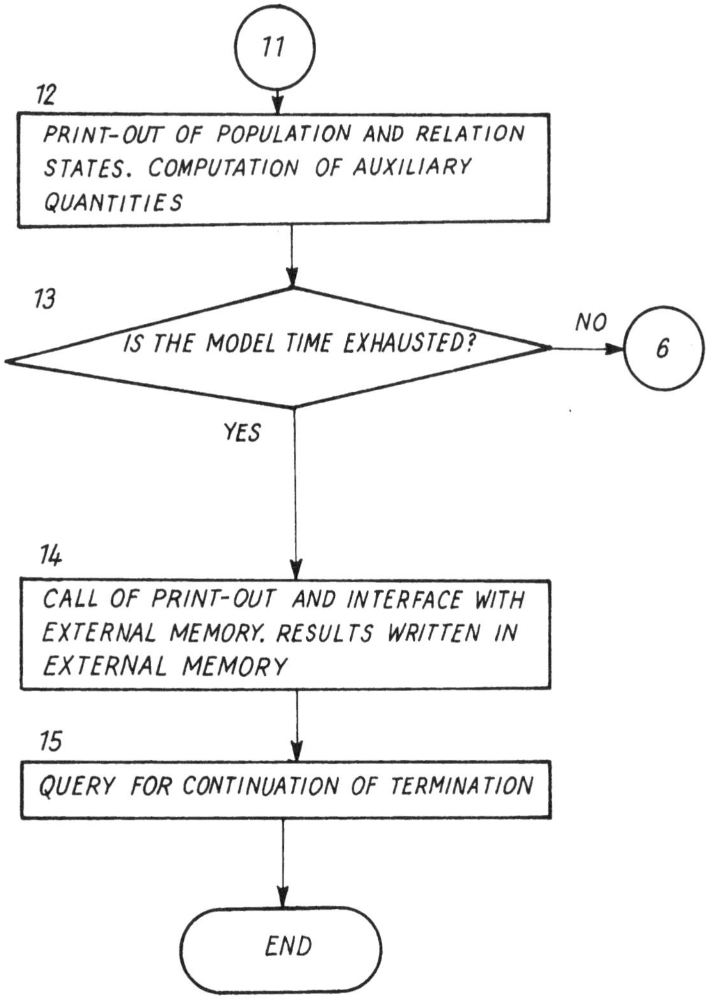

Fig. 31 (Continued)
OF MODULES FOR FINAL RECOMPUTATION OF POPULATION AND RELATION STATES AT THE END OF MODEL DAY. COMPUTATION OF NECESSARY QUANTITIES FOR SIMULATION OF COMMUNITY FUNCTIONING ON NEXT MODEL DAY (TIME T + ΔT); 12 - PRINT-OUT OF POPULATION AND RELATION STATES. COMPUTATION OF AUXILIARY QUANTITIES; 13 - IS MODEL TIME EXHAUSTED? 13a - NO, 13b - YES; 14 - CALL OF PRINT-OUT AND INTERFACE WITH EXTERNAL MEMORY. RESULTS WRITTEN IN EXTERNAL MEMORY; 15 - QUERY FOR CONTINUATION OR TERMINATION; 15a - END.

algorithms the sequential numbers of equations and relations in the mathematical description are cited in the Figures 31a and 32, Table 22.

Table 22 Quantities used in the mathematical description of interspecific relations and the associated equations of Chapter 2. Quantities listed as they appear in the discussion (Chapter 3)

Quantity, notation	Equation or Table
Hydroid	
P_H^*	2.1.6.
A_{POT}^*	2.1.7.
G_H^*	2.1.23.
r'^*	2.1.9.
N_E^*	2.1.11., 2.1.25., 2.1.30.
N_H^*	2.1.8., 2.1.26.
V_O^*	2.1.20., 2.1.25.
K_H^*	2.1.6., 2.1.8.
Ψ_H^*	2.1.6.
TR^*	2.1.20.
N_O^*	2.1.8.
C_W^*	2.1.1.
Balanus	
B_j^*	2.2.3.-2.2.5.
f_j^*	2.2.3.-2.2.8.
Ψ_B^*	2.2.3.
NS^*	2.2.8.-2.2.10., Table 3
NC^*	2.2.12.
Nudibranch	
BT_j^*	2.3.7.
b_j^*	2.3.3.
Infusorian	
n_p^*	2.3.8.
Crab	
BC_j^*	2.4.6.-2.4.9., Table 14

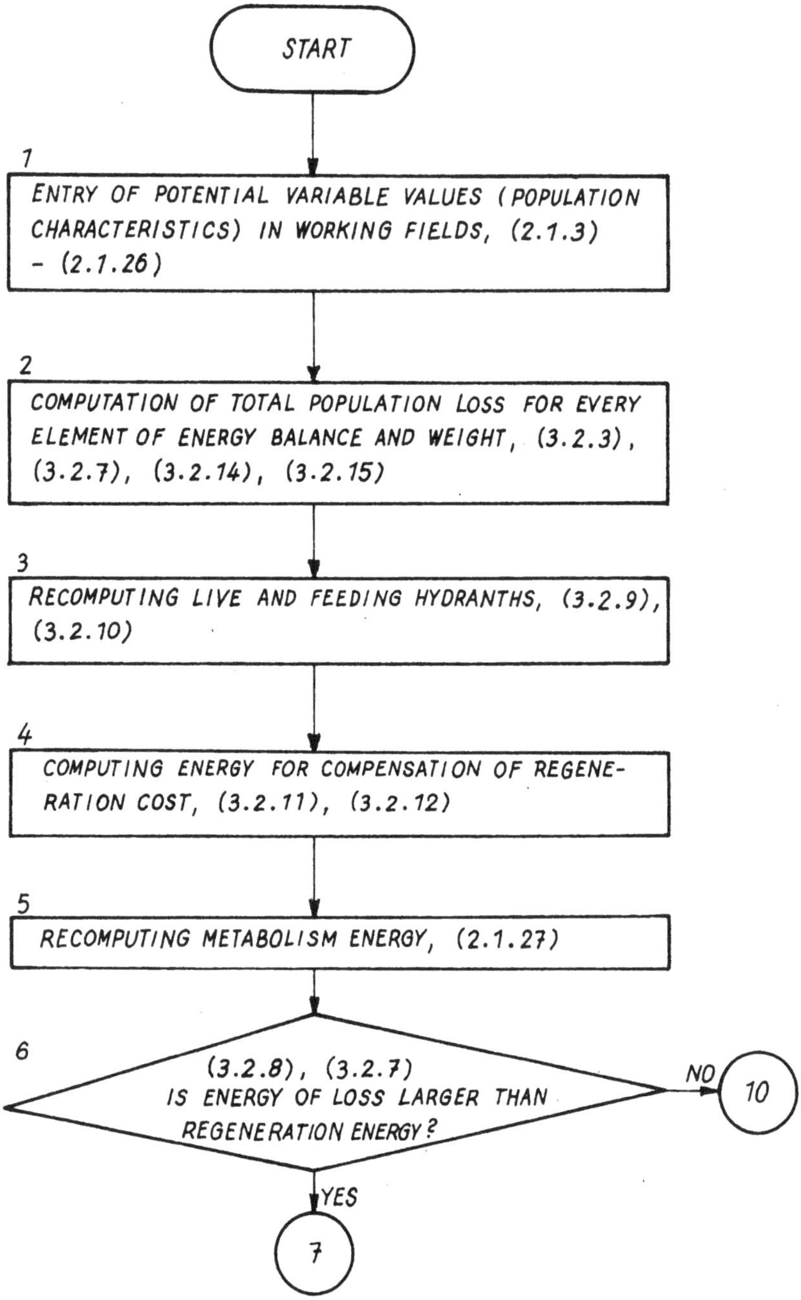

Fig. 31a Logical-heuristic algorithm of a module for recomputing the population state (hydroid).
0 - START; 1 - ENTRY OF POTENTIAL VARIABLE VALUES (POPULATION CHARACTERISTICS) IN WORKING FIELDS, (2.1.3.) - (2.1.26.); 2 - COMPUTATION OF TOTAL POPULATION LOSS FOR EVERY ELEMENT OF

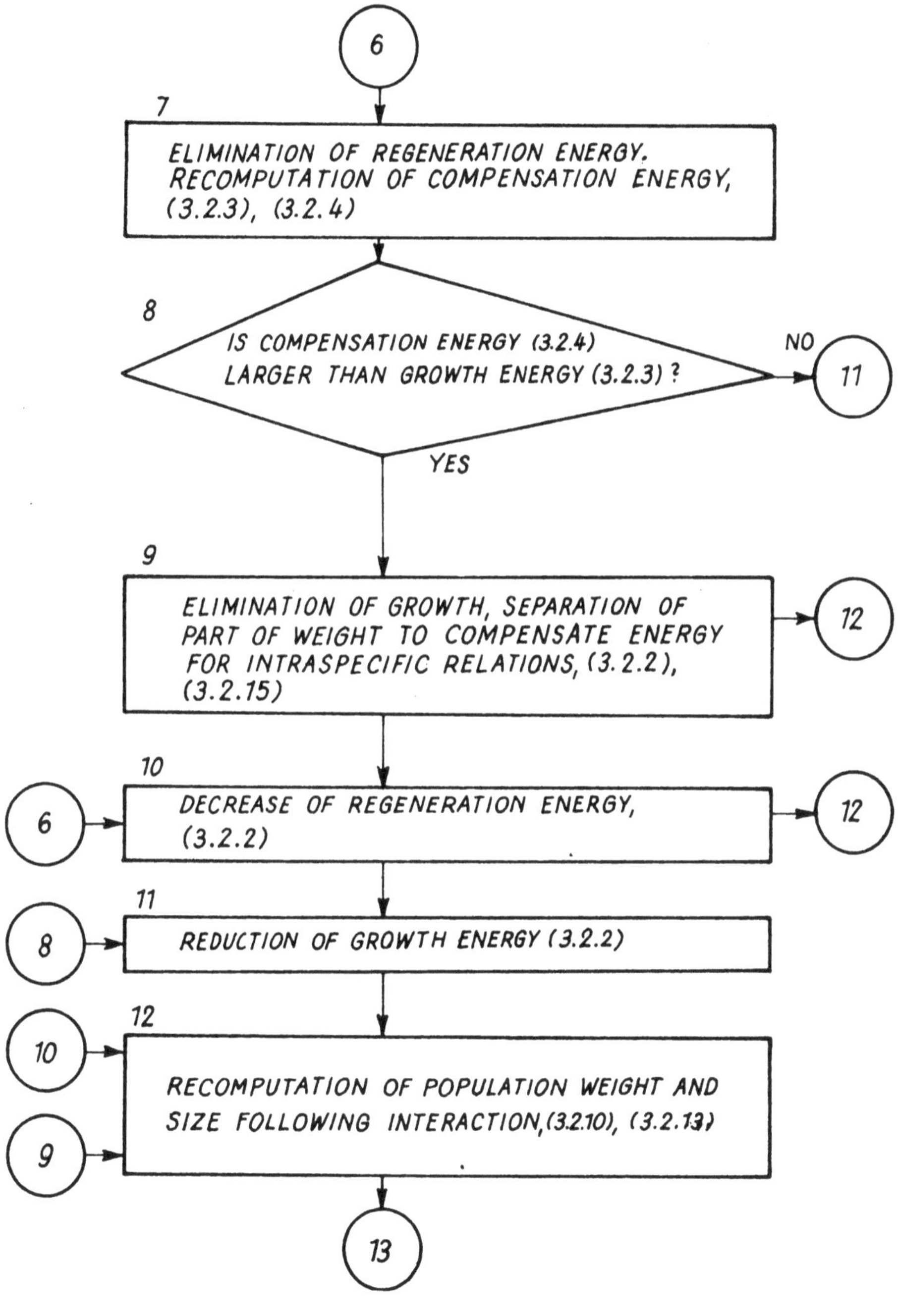

Fig. 31a (Continued)

ENERGY BALANCE AND WEIGHT, (3.2.3.), (3.2.7.), (3.2.14), (3.2.15.); 3 - RECOMPUTATING LIVE AND FEEDING HYDRANTHS, (3.2.9.), (3.2.10.); 4 - COMPUTING ENERGY FOR COMPENSATION OF REGENERATION COST, (3.2.11.), (3.2.12.); 5 - RECOMPUTING META-

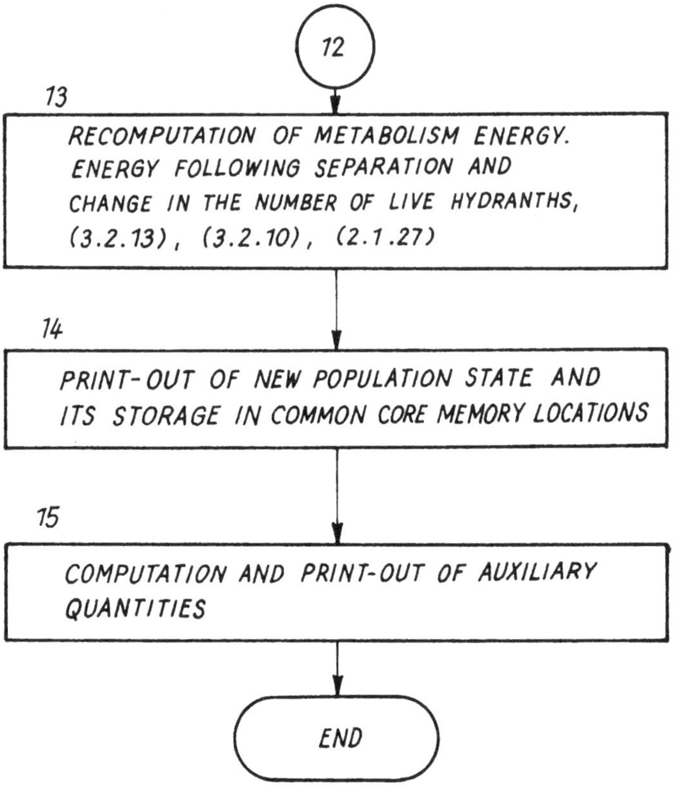

Fig. 31a (Continued)

BOLISM ENERGY, (2.1.12.); 6 - IS ENERGY OF LOSS LARGER THAN REGENERATION ENERGY, 6a - NO, 6b - YES; 7 - ELIMINATION OF REGENERATION ENERGY. RECOMPUTATION OF COMPENSATION ENERGY, (3.2.3.), (3.2.4.); 8 - IS COMPENSATION ENERGY (3.2.4.) LARGER THAN GROWTH ENERGY (3.2.3.)? 8a - NO, 8b - YES; 9 - ELIMINATION OF GROWTH. SEPARATION OF PART OF WEIGHT TO COMPENSATE ENERGY FOR INTRASPECIFIC RELATIONS, (3.2.2.), (3.2.5.); 10 - DECREASE OF REGENERATION ENERGY, (3.2.2.); 11 - REDUCTION OF GROWTH ENERGY, (3.2.2.); 12 - RECOMPUTATION OF POPULATION WEIGHT AND SIZE FOLLOWING INTERACTION, (3.2.10.), (3.2.13.); 12 - RECOMPUTATION OF METABOLISM ENERGY FOLLOWING SEPARATION AND CHANGE IN THE NUMBER OF LIVE HYDRANTHS, (3.2.12.), (3.2.10.), (2.1.27.); 14 - PRINT-OUT OF NEW POPULATION STATE AND ITS STORAGE IN COMMON CORE MEMORY LOCATIONS; 15 - COMPUTATION AND PRINT-OUT OF AUXILIARY QUANTITIES; 15a - END.

Software Characteristics

Before proceeding to the software proper, a few words are needed on the operating system (OS) of the IBM 360/370 family of computers. The OS handles all available resources including the central processor, main memory, the input/output facility, and service programs, which are part of the OS (Madnik and Donavan, 1974). The necessary information is indicated in a dedicated task control language which interfaces the user's processing program and the OS. This function is performed by control cards, essential for OS scheduling components and used for allocating the system resources and scheduling. This flexible tool can be considered a high level programming language. Since most tasks do not require all of the OS potential, the task scheduling by this language may be simplified by the use of libraries of standard sizes and the development of special-purpose software (Madnik and Donovan, 1974; Davis, 1977). An attempt to develop such software was made in developing a simulation model for the fouling community. A library of modules for autonomous models was made possible by a unified structure of the population and of the model itself (Sections 1.2. and 1.4.). The modules called by the head module of the simulation program are formally programmed as parametrically independent; in other words, the formal parameters were in fact replaced by using standard data structures which do not depend on the parameter value. The simulation program was controlled by the head module, where the model adjustment cards transmitted data on model simulation options. All modules of the simulation program were identified by the functional feature. The efforts invested in the multistaged development of the model over years made the simulation program fairly modifiable and flexible. The overall structure of the entire software for the fouling community is shown in Figure 33.

Because the quantitative characteristics of simulation programs for models of an autonomous population were discussed in Chapter 2, this Section will describe only the simulation program for the community model. It consisted of 15 basic modules (other than the software of autonomous models) and four auxiliary modules. Table 23 lists quantitative characteristics of the modules. The total time of a full program run was 42 minutes, the CPU time was 18.2 minutes. When the core memory and the run time were limited, an abridged modular option was used whose run time was 24 minutes and the CPU time, 6.85 minutes. There were 420 basic variables in the simulation program. In the complete modular option there were 2,950 FORTRAN statements, an execution step required a 332 R core memory and the abridged option, 1,492 statements and a 290 K core memory.

Simulation design. The generation, development, and functioning of the fouling community depend on the environment, interaction of the

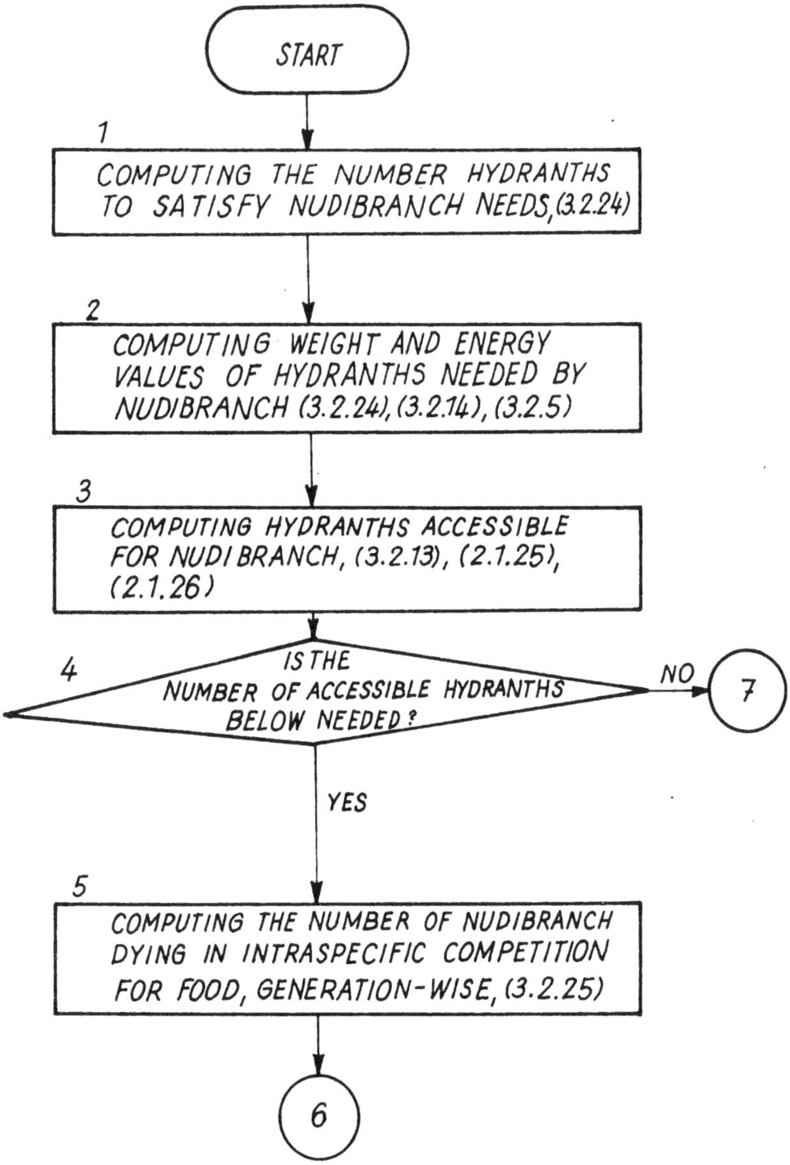

Fig. 32 Logical-heuristic algorithm of the module for recomputing the state of relation No. 3 (nudibranch - hydroid)
0 - START; 1 - COMPUTING THE NUMBER OF HYDRANTHS TO SATISFY NUDIBRANCH NEEDS, (3.2.24.); 2 - COMPUTING WEIGHT AND ENERGY VALUES OF HYDRANTHS NEEDED BY NUDIBRANCH, (3.2.24.), (3.2.14.) (3.2.5.); 3 - COMPUTING HYDRANTHS ACCESSIBLE FOR NUDIBRANCH, (3.2.13.), (2.1.25.), (2.1.26.); 4 - IS THE NUMBER OF ACCESSIBLE HYDRANTHS BELOW NEED; 4a - NO, 4b - YES; 5 - COMPUTING THE NUMBER OF NUDIBRANCHES DYING IN INTRASPECIFIC COMPETITION

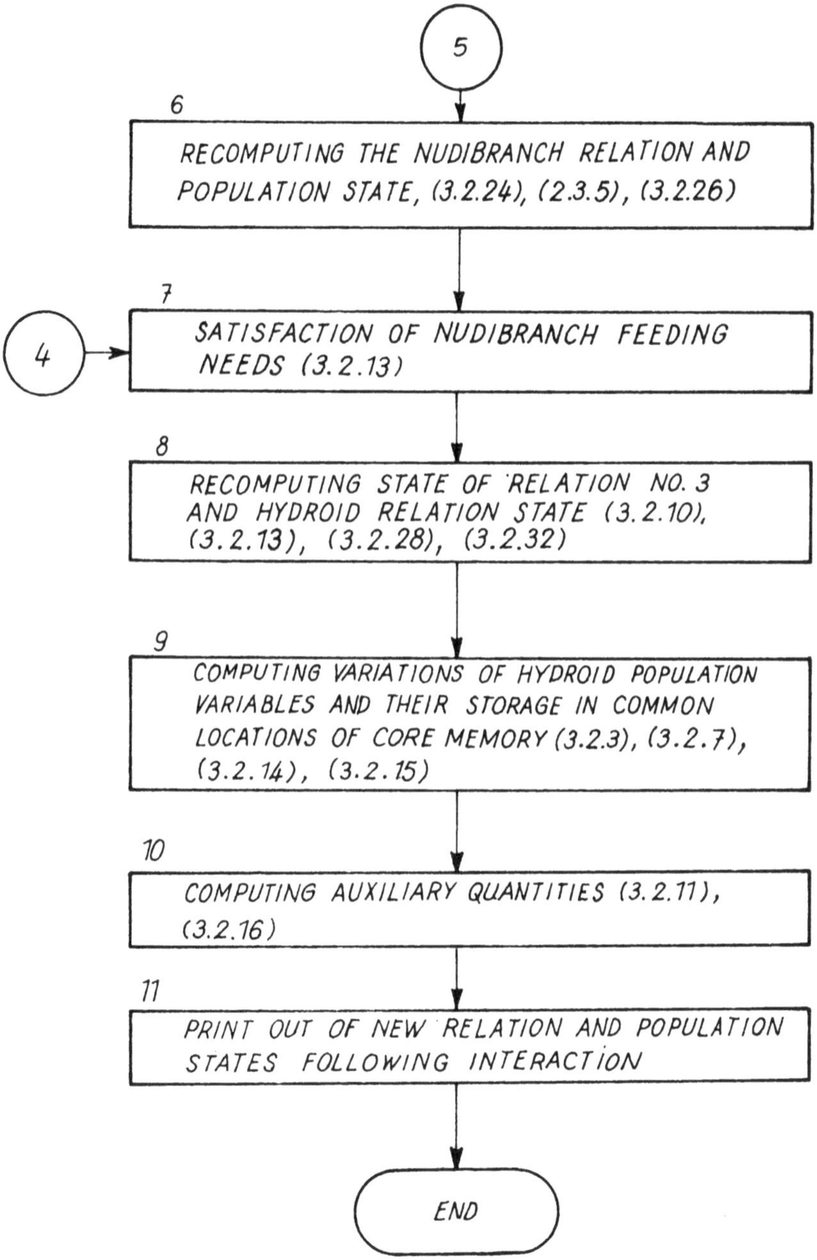

Fig. 32 (Continued)

FOR FOOD, GENERATION-WISE, (3.2.25.); 6 - RECOMPUTING THE NUDIBRANCH RELATION AND POPULATION STATE, (3.2.24.), (2.3.5.), (3.2.26.); 7 - SATISFACTION OF NUDIBRANCH FEEDING NEEDS; 8 - RECOMPUTING STATE OF RELATION NO. 3 AND HYDROID RELATION STATE, (3.2.10.), (3.2.13.), (3.2.28.), (3.2.32.); 9 - COMPUTING VARIATIONS OF HYDROID POPULATION VARIABLES AND THEIR STORAGE IN

COMMON LOCATIONS OF CORE MEMORY (3.2.3.), (3.2.7.), (3.2.14.), (3.2.15.); 10 - COMPUTING AUXILIARY QUANTITIES (3.2.11.), (3.2.16.); 11 - PRINT-OUT OF NEW RELATION AND POPULATION STATES FOLLOWING INTERACTION; 11a - END.

populations, the food supply and abiotic factors. The model community was investigated in a series of pre-designed experiments which represented various possible or observed conditions. The dynamics of basic population characteristics in every series (or for specific conditions) of the simulation experiment was forecasted in several model runs with different food concentrations (in thousands of phyto- and zooplankton specimens per cubic meter of water) and abiotic factors (salinity of five and eight permille and temperature as a signaling factor).

The development of the community depends chiefly on the hydroid polyp, which grows as a natural substrate for other main fouling species such as the colonial infusorian (food species) the nudibranch (first-order predator), and crab (second-order predator). The nudibranch competes with the hydroid for the artificial substrate. Adult nudibranches consume the hydroid hydranths, and adult crabs feed on adult nudibranches. The mechanisms and the time of interaction may vary.

Table 24 summarizes the qualitative characteristics of different series of the simulation experiment. The specification of parameters in the simulation experiment for different conditions formation and functioning of the community made it possible to estimate the impact of certain factors on the biosystem. The quantitative parameters of runs for every option of the model (simulation series) are presented in Table 25. The adequacy is analyzed for the series IV, whose conditions represent those of a field experiment.

3.4. Adequacy Estimation of the Simulation Model

All models were analyzed, Chapter 1, by comparing the simulation results and the available field and ecological data in quantitative and qualitative terms. For simulation models which faithfully represent the basic processes in the biosystem, qualitative consistency is more important than quantitative fit in adequacy analysis.

In the spring and summer period covered by the model the community functioning was dictated by hydroid - balanus and infusorian - hydroid relations. The field data and the experimental type of the formation and development of the community were represented by the series IV (for the single-layer balanus) and the series V (for the multi-layer balanus) in the model of Section 3.3. Therefore, the adequacy was checked against the quantitative simulation results obtained in those series.

Table 23 Quantitative characteristics of the main simulation program for the community model

Programming module	Number of FORTRAN statements	Core memory, Kbytes, of object module
Head	135	8
Hydroid state recomputation	109	4
Balanus state recomputation	38	1.5
Nudibranch state recomputation	112	5
Infusorian state recomputation	42	1.5
State of relation No. 1	44	1.5
State of relation No. 2	51	2
State of relation No. 3	160	7
State of relation No. 4	119	5
State of relation No. 5	57	1.5
State of relation No. 6	107	4.5
State of relation No. 7	61	2
State of relation No. 8	99	2.5
State of relation No. 10	23	1
Auxiliary controls	31	1
Input/Output	68	1.5
Plotting	129	2.5
General memory	37	1
Entire program	1,491	55

Models of autonomous populations were shown to be adequate in Chapter 2; for every population of widespread species numerous simulation experiments were carried out in model conditions for which control data were available. Models of autonomous populations were found to provide results which agree with both quantitative field data and qualitative ideas on basic processes in the functioning of populations (Turpayeva and Gal'perin, 1980 a & b; Gal'perin et al., 1981; Turpayeva et al., 1981; Gal'perin and Turpayeva, 1982). Consequently, the adequacy of the community model can be considered the adequacy of quantitative and qualitative results obtained from the simulation of processes in the species' interaction. This analysis can be carried out for basic complexes of relations.

The control data and computed characteristics of the hydroid - balanus relation for a single-layer balanus settlement are shown in Fig. 34. The man growth rate of the balanus $\overline{C}_B$ as a function of that of the hydroid $\overline{C}_H$ is nonlinear. The hydroid does practically not inhibit the balanus if $\overline{C}_h < 0.4$, while this value is 0.3 for the control curve. Although the pattern of curves for computed and control data is

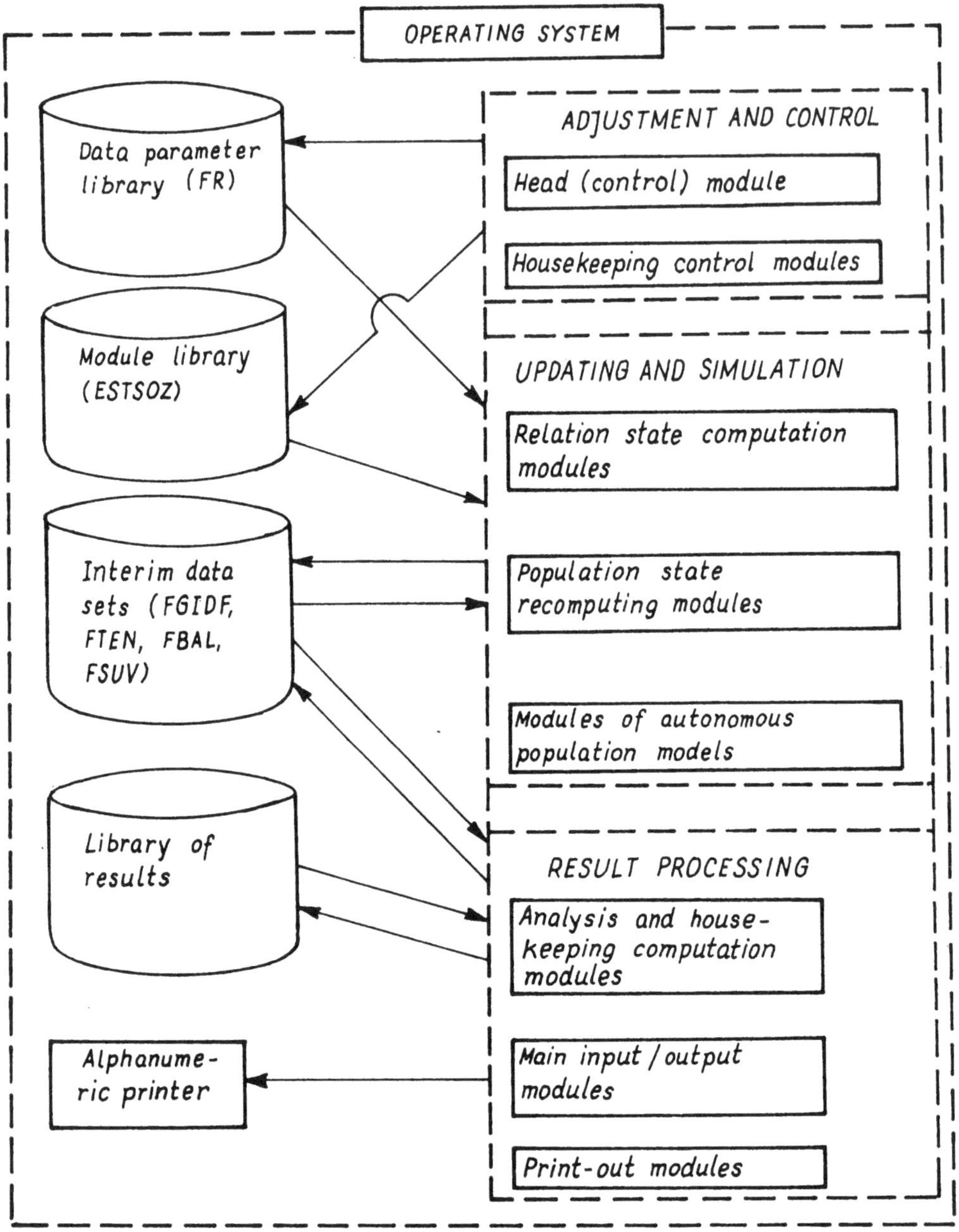

Fig. 33 Overall software structure for the simulation model of a fouling community

1 - Operating system. 2 - Data parameter library (FR). 3 - Adjustment and control. 4 - Head (control) module. 5 - Housekeeping control modules. 6 - Module library (ESTSOZ). 7 - Updating and simulation. 8 - Relation state computation modules. 9 - Interim data sets (FGIDT), FTEN, FBAL, FSUV). 10 - Population state recomputing modules. 11 - Modules of autonomous

population models. 12 - Library of results. 13 - Result processing. 14 - Analysis and housekeeping computation modules. 15 - Main input/output modules. 16 - Alphanumeric printer. 17 - Print-out modules.

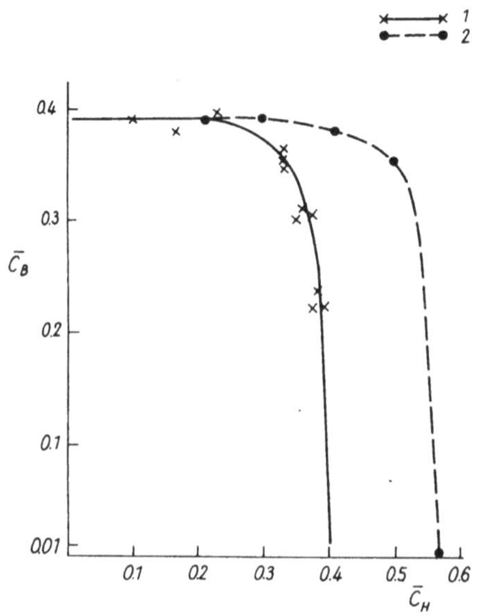

Fig. 34 The average balanus growth rate $\bar{C}_B$ versus the average hydroid growth rate, $\bar{C}_h$.
1 - Control data. 2 - Computed data.

identical, there is a short delay in the development of the balanus pressed by the hydroid because, when the control data were collected, the balanus on the plates were given three meals a day rather then fed continuously as in an actual community (Turpayeva, 1976) and the simulation model reflects the latter situation.

An increase of the hydroid growth rate to 0.5 dramatically inhibited the growth of the balanus population, which was barely adult by the end of the life time. A further increase of β caused the growth and development of the balanus to cease (in different runs $\alpha = 120$; $\beta = 100$; $\alpha = 120$; $\beta = 50$). In the runs of the fourth series $\alpha = 70$, $\beta = 20$ and $\alpha = 50$, $\beta = 10$, and the barnacles lived to the adult state. If $\alpha = 170$ and $\beta = 5$, the balanus developed normally. If $\alpha = 30$ and $\beta = 5$, with either salinity level (five and eight permille), both populations were underdeveloped. The metabolism energy accounted for 88 to 92 per cent of the energy flow in the inhibited balanus.

A comparison of simulation and control data for the hydroid - balanus relation (No. 8), if the barnacles make a multi-layer population,

Table 24 Comparative qualitative description of simulation experience series (simulation design)

Series No. Population	I	II	III	IV	V
HYDROID	Population runs of zooplankton on 80th model day temperature change	Likewise	Food available during entire model period	Likewise	Likewise
BALANUS	Single-stratum settlement	Likewise	Likewise	Likewise	Multistratum settlement
NUDIBRANCH	Continuously feeds on infusorian. All adults lay eggs before dying	Likewise	Leaves damaged areas to allow infusorian to restore. Thirty per cent of adults lay eggs before dying	As II and III but per cent of adults lay eggs before dying	Likewise
INFUSORIAN	Non-regenerated part of hydroid colony is substrate. Restoration of consumed areas in a day or two	Regenerate part of hydroid colony is substrate. Restoration in two or three days	Newly growing part of hydroid colony is substrate (about 20 per cent of generative part). Restoration as II	As IV	As V
CRAB	Feeds only on infusorian and nudibranch. Cannibalism not pronounced	Likewise, but cannibalism by 16 per cent higher	Crab's diet includes nudibranch eggs	Juvenile crabs feed initially accessible egg deposits	As IV

Table 25 Quantitative characteristics of model runs for every simulation option (simulation design)

Parameter	Series I - IV												Series V		
	1	2	3	4	5	6	7	8	9	10	11	12	13	14	15
Phytoplankton concentration (α, thousands/ m^3)	120	120	120	120	120	120	170	70	50	120	30	30	100	120	100
Zooplankton concentration (β, thousands/ m^3)	100	100	50	30	30	10	5	20	10	40	5	5	40	100	100
Salinity (S, per mille)	8	5	8	8	5	5	8	8	8	8	8	5	8	8	8

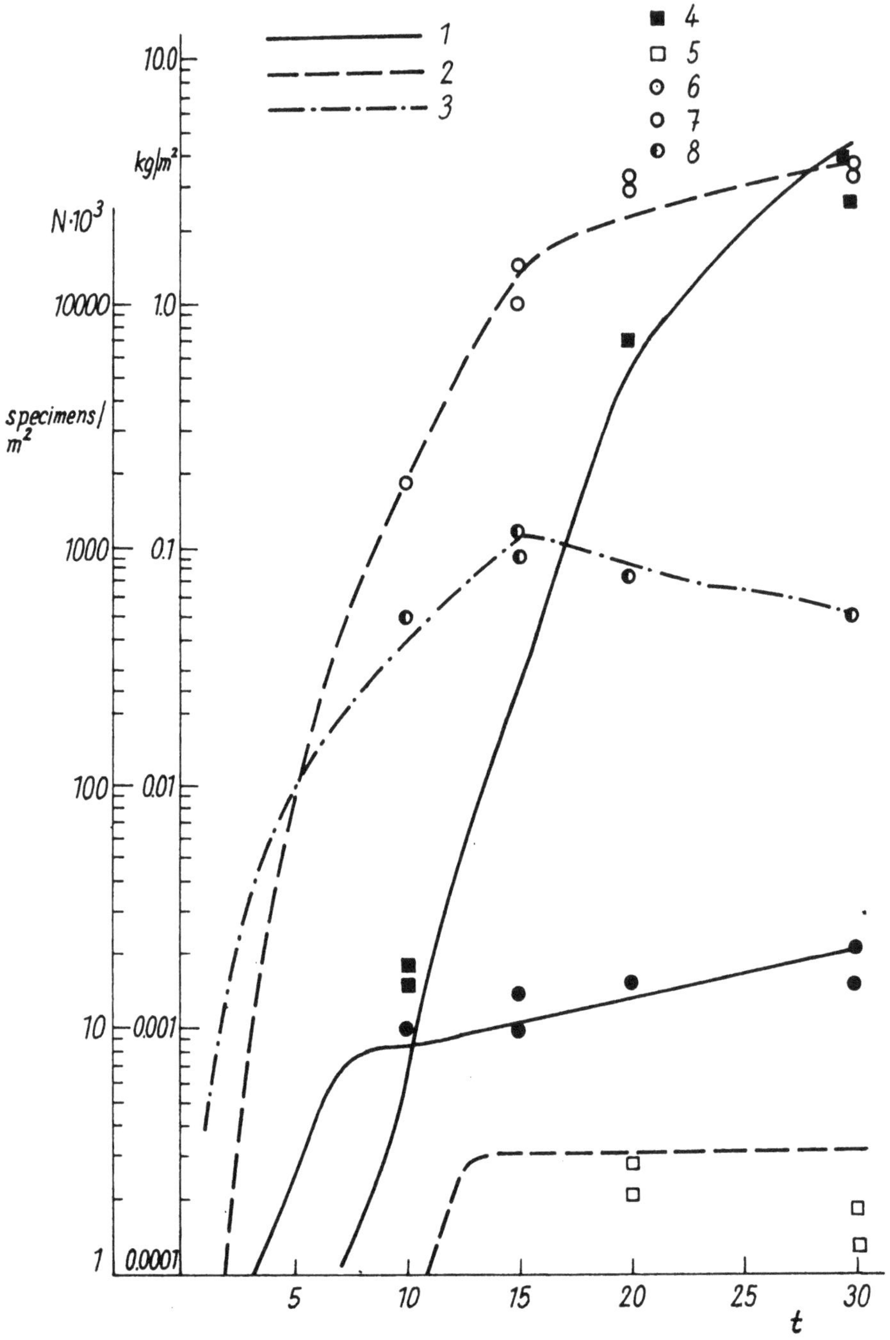

Fig. 35 The hydroid biomass versus the number of barnacles making a multi-stratum settlement.
1, 2, 3 - computed data. 4 and 5 control data (1968); 6 - 8 control (1969). The hydroid biomass (1), balanus biomass (2), number of barnacles in the population (3).

is shown in Fig. 35, where the curves represent three characteristic options of barnacle settlement, Section 2.2. As the number of settled larvae increases, the hydroid is inhibited. The time when this inhibition starts begins somewhat "later" than the strong intraspecific competition develops, curves 1 and 2. The computed curves go practically through all control points which were not used in the computation. Before the inhibition starts, the hydroid parts grow exponentially allowing an extended period of the settlement and development of original polyps. When the balanus occupies a large substrate area, the growth of the hydroid falls dramatically to 0.002, 0.014, and 1.63 kg/m^2 in various options, which is seen in Fig. 35 to coincide with the control. When the settled larvae are abundant, the strength of the relation initially increases and, following the inhibition, remains at approximately the sáme level. The relation No. 8 is the strongest when the two populations actively compete for the substrate and is in some cases three to five times stronger than that, when one of the populations is inhibited. The relation No. 8 varies in a fairly wide range when the direct trophic nudibranch - hydroid relation (No. 3) intensifies.

The ratio of computed and control values for the latter relation is shown in Fig. 36. To illustrate the mechanism of this relation, the Figure plots the indicator of the hydroid nutrition regulation (computed as the ratio of potential to actual consumption of assimilated food) as a function of the density N_u of the nudibranch population, which was expressed as the average number of hydranths in the area taken up by one mollusk. The relation No. 3 could not be expressed in terms of the growth rate because the populations start interacting on the 36th day of the model period when the hydroid feeding regulation becomes strong. This relation is most intensive in the modelled flowchart of synphysiological relations. The curve of Fig. 36 is logarithmic; even for relatively low (100 - 125) N_u, δ is maximal. With $N_u \leqslant 23 - 25$ the nudibranch does not practically inhibit the hydroid. Inhibition occurs if β is low, 1 - 20, because the hydroid is underdeveloped, and so the nudibranch population does not grow because of the juveniles dying on the infusorian. As $N_u > 125$ is very rare in real life, the curve is traced in compliance with the overall trend.

Two direct trophic relations, nudibranch - infusorian and crab - infusorian, dictate the development of juveniles in the populations of conditioning species.

By consuming it the nudibranch stimulates the development of the infusorian; on the other hand, when food is in short supply, the nudibranch juveniles die and the survivors move to intact parts of the colonies. As an omnivorous species the crab is less dependent on the infusorian. The quantitative characteristics of these two relations

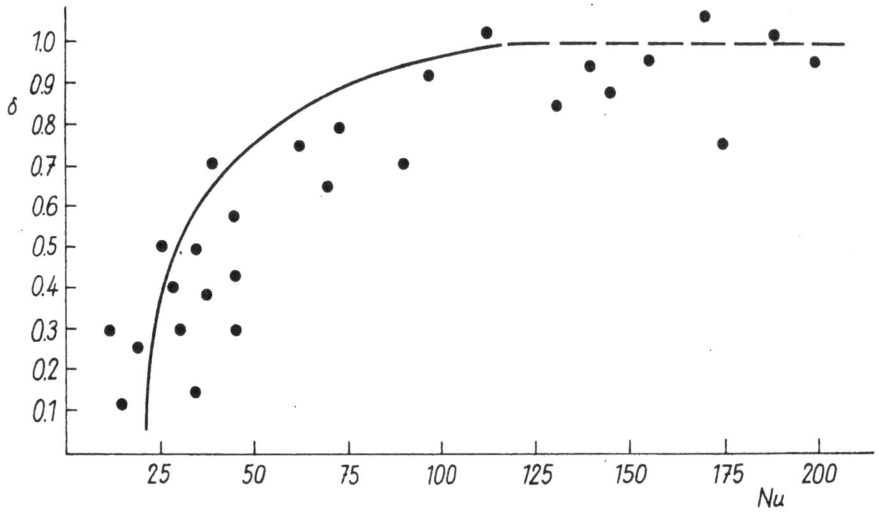

Fig. 36 The hydroid regulation indicator ($\delta = A'_H/A_H$, Section 2.1.) as a function of the nudibranch population density (N_u). The solid line represents control and the dot line computed data for various runs. (Turpayeva, 1982)

were dictated by the fractions of the infusorian in the daily ration of nudibranch and crab juveniles. Since the algorithms of program modules were consistent with the mechanisms of natural processes, the quantitative characteristics of fractions in the diet were taken from field data, and the computed strengths of trophic relations agree with the experimental findings, the results of simulating the relations Nos. 4 and 7 could be accepted as adequate (Turpayeva, 1976; Gal'perin et al., 1981; Gal'perin and Turpayeva, 1982; Turpayeva, 1982; Gal'perin, 1983 a–e).

Consequently, the good fit of the computed and empirical points for control options of simulation seems to confirm the adequacy of the simulation model. Furthermore, the analysis of the simulation results for the consistency of the qualitative model functioning and the actual system tend to confirm the qualitative adequacy of the behaviour of the model and the object. In the same way as the autonomous models were found adequate in Chapter 2, a conclusion can now be made that the results of the simulation of the entire fouling community are adequate.

3.5. Statistical Significance of the Results and Analysis of the Impact Made by the Computation Method

The evaluation and adequacy analysis of the simulation results must be supplemented by processing basic quantities used in the adequacy analysis for statistical significance. Let us assume that the computation results of the Figures 34 - 36 were obtained from a regression equation. The significance of a regression equation may be computed in terms of the Fisher criterion F (Brandt, 1970; Afifi and Eisen, 1979). The general variance $\bar{S}^2$ is compared with the residual variance $\bar{S}^2_{res}$ by the formulae of Section 2.5. The residual variance indicates the error in predicting the experimental results by the regression equation. In our case this is the simulation error, the deviation from available field data which were not used in developing the model. The statistical analysis confirms or disproves the significance of the computing method, which was dictated, in this case, by biological considerations of the community model as well as of this curve. The prediction performance may be determined by comparing the general and residual variances. The Fisher criterion shows the significance of this kind of regression, or the ratio of the results predicted by the regression equation to the mean value. In our case, this implies the significance of parameter variations and simulation runs compared with some "averaged" run using constant parameters.

For the hydroid - balanus relation of Fig. 34 the general variance $\bar{S}^2$ was found to be 0.111, and $\bar{S}^2_{res}$ 0.024. In this case, the significance of the simulation results is one per cent, the tabular value was 4.539 while the actual value of the F^T-criterion was equal to 4.583.

For the nudibranch - hydroid relation the computation showed that the significance was at the 2.5 per cent level. With 27 test points (Fig. 36), $\bar{S}^2 = 0.121$, $\bar{S}^2_{res} = 0.048$, $\hat{F} = 2.513$, and $F^T = 2.242$. An analysis of the residues is made to prove the applicability of the statistical treatment to any regression dependences (Brandt, 1970; Afifi and Eisen, 1979). In regression analysis the errors are assumed to be independent, having zero means, constant variance, and they would be normally distributed. First, the mean error is computed. For the nudibranch - hydroid relation it is - 0.0444, or 4.5 per cent of the average. The sign "-" implies that the observation results are above the curve on the average. The scatter range over the variance is 4.02. The bounderies of the confidence interval in the Table of (Brandt, 1970) are 3.45 and 4.53, resp. Consequently, the errors may be assumed to be normally distributed.

The significance of the results for the balanus - hydroid relation, Fig. 35, was found to be at the 10 per cent level, although there were only a few test points.

The impact made by the computation method. In implementing the

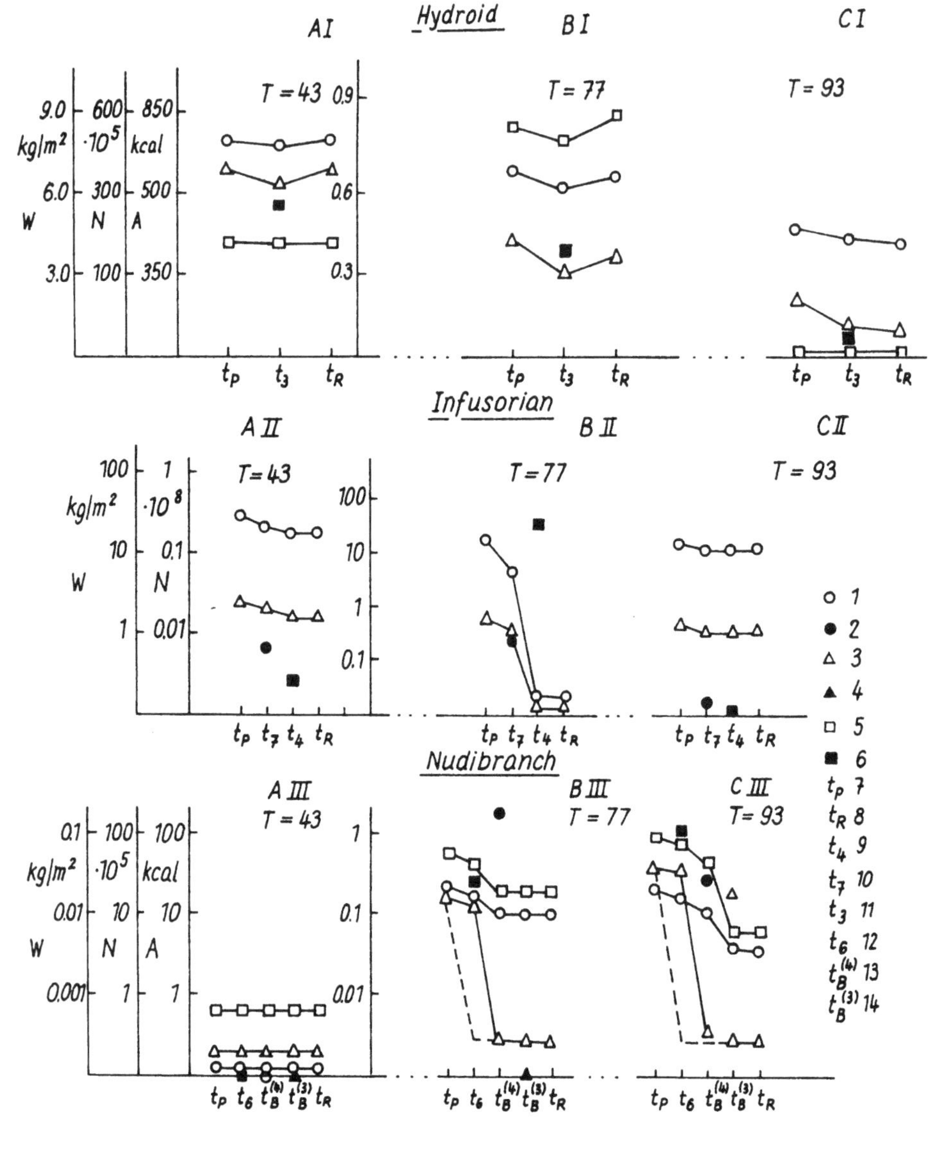

Fig. 37 The dynamics and nature of the functioning of the species interaction system in the community within one simulation step. 1, 3, 5 - weight, size, and assimilated food of populations. 2, 4, 6 - relative rates of interspecific relations at the time of mode implementation. 7, 8 - time of computing the potential and actual values of basic characteristics of the populations. 9 - 12 - times of implementing models of interspecific relations (Sequential numbers of modules of Table 23) Nos. 4, 7, 3, 6, and (13, 14) intraspecific competition of the nudibranch on the infusorian and hydroid.

general simulation algorithm, which reflected the model concept and research methodology, two assumptions were made on the priorities of relations. Firstly, topic relations were implemented first. Second, relations active over the longest periods had a higher priority. These assumptions proved to be sufficient for quantifying the process and computing the sequence of calling the associated modules. It was, however, necessary to assess the impact made by this method on the simulation results. In Fig. 37 the solid and dotted lines connect the values obtained for the implementation sequence used in the model and another possible sequence which is obtained by the assumption that interspecific relations of higher potential rate have a higher priority. Figure 37 plots the results of a simulation run of the model with a salinity S = 8 o/oo and a zooplankton concentration β = 50 (thousand specimens per cubic meter of water).

The days subjected to analysis are taken from characteristic periods in the development of the model community. On T = 43, the community is at the initial stage of its development. The nudibranch predator population produces no tangible effect on the potential of dependent populations although it reduces the rate of intra-population hydroid nutrition (parts AI, AII, and AIII in the Figure). On that model day the potential and actual characteristics of interacting populations nearly coincide. When the sequence of implementing the modules is changed, the computation results do not significantly change.

On T = 77, the situation essentially changes in that the interaction of the most intensive specimens, in terms of kg, and kcal, of the species starts (Fig. 37 BI, BII, and BIII). The infusorian is inhibited. Consumed by the crab and nudibranch, the infusorian biomass falls by about two orders of magnitude, from 0.16 to 0.002 kg. The predators also inhibit complete restoration of this population. The nudibranch - hydroid relation makes an impact on the dependent hydroid species, which is stronger than that on T = 43, for its biomass is about 1 kg below the value of the autonomous model while the assimilated food is larger (633 kcal in this run against 413 kcal in the autonomous model). This occurs by restructuring the contributions made by elements to the hydroid energy balance, and by the action of compensatory mechanisms. The feeding regulation indicator is δ = 1, Section 2.1., which implies that all living hydranths have food available and make maximum use of it. The intraspecific competition of juvenile nudibranches proves stiff because of shortage of the infusorian. Nudibranch adults still have food in good supply because the hydroid polyp has preserved enough hydranths. For this reason the death of most of the juvenile predators does not much reduce the weight and assimilated food of the population. The relative rates of interspecific relations dramatically increase.

On T = 93 the community is inhibited (Fig. 37 CI, CII, and CIII).

No food is available for the hydroid, but there are enough hydranths to keep some of the nudibranches alive. Nearly all nudibranch juveniles die because the infusorian is inhibited and is not even partially restored. The contribution made by the crab, whose large specimens consume adult nudibranches, has increased four-fold over the preceding period. Parts of the hydroid colony for which food is insufficient for the energy metabolism break away at a high rate. The topic relations are noticeably weaker.

For the simulation results obtained on T = 77 and T = 93 the change in the sequence of implementing the models of interspecific relations plays a significant role in case of both the nudibranch and crab consuming the infusorian. When the sequence changes, the absolute value of the intraspecific competition in the nudibranch population decreases, because of shortage of infusorian heads, by an average of 15 - 28 per cent, which is equal to the infusorian fraction in the crab's diet. This difference occurs only in runs of different series where the infusorian is in an unfavourable state because the hydroid colony is underdeveloped. In most series these runs represent the cases of low plankton concentration (β = 10), and in the series where the hydroid is fully developed but the substrate area for the infusorian is small, the case of high plankton concentration (β = 30). In the run shown in Fig. 37 the difference in the intraspecific competition of juvenile nudibranch amounts on the average to eight to 10 per cent of the energy value stored in the dying nudibranches over the model period when the relations of the nudibranch - crab, nudibranch - infusorian, and crab - infusorian relations compete. On the other hand, the range of basic characteristics of, say, the nudibranch population, is approximately 100 per cent of the values on T = 43. The range of simulated basic characteristics of the crab population averaged 25 to 30 per cent, and that of the hydroid 50 to 60 per cent of the maximum value. The impact of a change in the sequence of implementing the models on the rate of the crab - nudibranch and nudibranch - hydroid relations may be estimated in terms of the nudibranch biomass as 10 to 20 per cent, and in terms of the energy value, as 10 to 15 per cent, with the daily food assimilated by adults being 100 per cent. On the days when some adult nudibranches die because of a shortage of living hydranths, the difference for intraspecific competion rates amounts to 10 to 12 per cent of the total number (the energy value) of adult nudibranches.

Consequently, the comparison of different runs of the simulation model in terms of the dynamics of basic population characteristics suggests that a changed sequence of implementing the models of interspecific relations results in some differences in estimates of interspecific relation rates, but leaves the values of the characteristics practically intact. This difference is maximum for estimates of the intraspecific competition of the nudibranch population, which competes with the crab for food while some adult nudibranches fall prey to the crab.

The pattern and dynamics of the interspecific relations in a model fouling community seem to shed light on the mechanism of symphysiological relations. The proposed approach to the development of a simulation model (Section 1.4.) made it possible to simulate the situation in one simulation step. This approach is sound, in particular, because the impact of the computing technique on the numerical results is intelligible. In most simulation options the numerical results were found consistent biologically; as far as accuracy is concerned, Section 1.3., the results are sensitive to, for instance, changes in the sequence of implementing models of interspecific relations. Indeed, for more involved cases such as simultaneous mutual consumption, a correction is necessary to allow for this kind of changes, too.

3.6. Dynamic Forecasting of System Functioning Modes and of Basic Community Characteristics for Various Model Functioning Conditions

The Figures 38 and 39 plot the dynamics of the variations of the biomass and energy flow for the dominant species, i. e. the hydroid, in the community and the topically dependent populations in different functioning conditions, Table 24. Let us analyze thoroughly the cases of most different conditions in the simulation series I and IV for $\beta = 50$ and S = 8 permille.

Before the mobile nudibranch predator arrives and develops in the community (T = 36) the plots of characteristics in the series I and IV correspond to each other. The hydroid develops at a high rate and is capable of increasing the amount of assimilated food by 87 per cent. In the same period the numerical values of energy flows coincide for populations which are in topic dependence on the hydroid. Thus, when the hydroid biomass on T = 36 is about 7.5 kg, the crab biomass is 0.06 kg, the nudibranch biomass 10^{-4} kg, and the infusorian biomass 0.16 kg. As the nudibranch population develops in the first days of T = 36 - 43, the computed hydroid biomass is larger than in the autonomous model (curves 1, 3, 5, Fig. 38), which can be attributed to a partial elimination of the nutrition regulation ($\delta \approx 0.17 - 0.18$). By T = 44, the nudibranch consumes a lot of the hydranths, and the regeneration needs compel the hydroid population to resort to compensatory mechanisms, Section 3.2. The hydroid energy flow increases dramatically, the growth rate falls, and δ increases abruptly. The infusorian meets strong pressure from the nudibranch and in the series IV (Fig. 39 E), the biomasses of both species vary because the infusorian consumption and regeneration periods alternate. In the series I, the nudibranch population grows

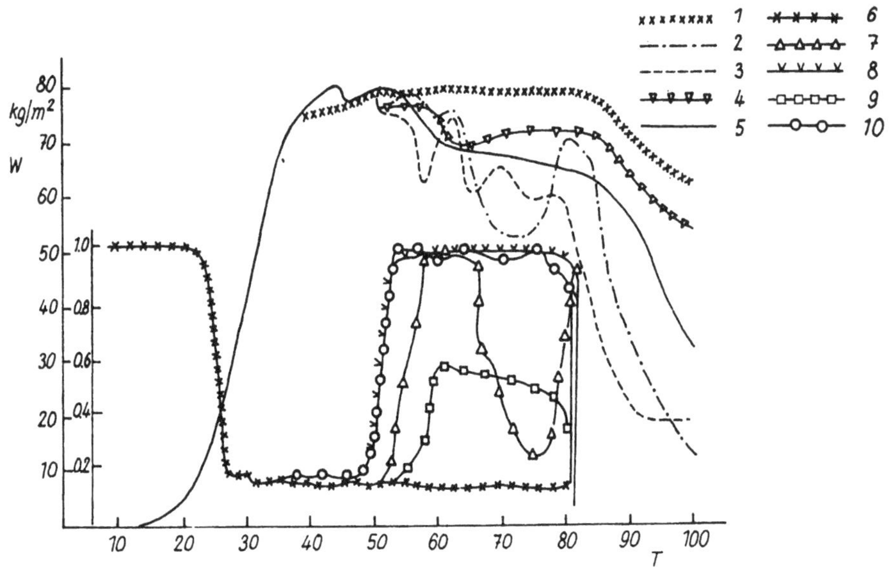

Fig. 38 The dynamic forecast of the biomass and the feeding regulation indicator δ of the hydroid polyp for various model functioning conditions (see Tables 24 and 25).
1, 6 - the biomass and the feeding regulation indicator from simulation results in the autonomous model. 2, 7 - the same for series II of simulation. 3, 8 - the same for series I. 4, 9 - the same for series III. 5, 10-the same for series IV.

fast, the biomass being approximately twice that of the series IV. Within 10 days the predator inhibits the infusorian and keeps it inhibited until nearly all juvenile nudibranches die (T = 72 - 78). The infusorian regenerates fast but is again inhibited because of the dwindling hydroid biomass and consumption by the crab, Fig. 39 D. The fast development of the infusorian in the first days of the series I makes it possible for a larger number of adult nudibranches to survive in the population by the time when the intraspecific competition increases, rather than in the series IV, where the age structure of the population is more balanced. In the series I, adult nudibranches are favoured at every time while the juveniles are sacrificed. In the mechanism represented by curve 3, Fig. 38, the high rate of the nudibranch population development enables numerous adults to reduce the hydroid population drastically. As a result, some nudibranches die of food shortage. The hydroid is capable of regenerating for a short time. As the hydroid develops, the infusorian restores. Juveniles hatch from eggs deposited by the now dead nudibranch adults. Some of these animals grow to large sizes and feed to an increasing extent on the hydroid. The process resumes. In the series IV we meet the slow development of the juveniles

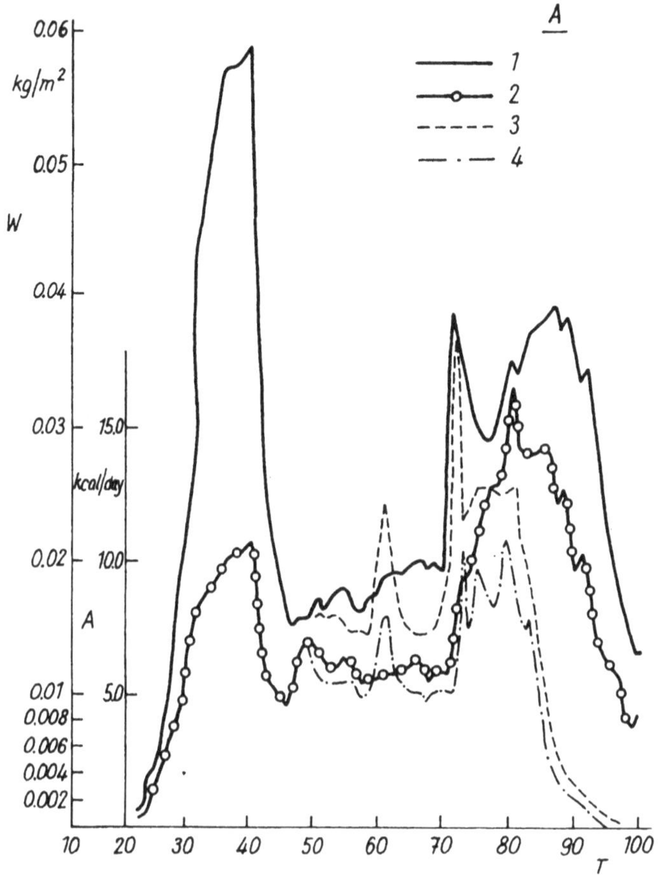

Fig. 39 A. Biomass (1, 3) and assimilated food (2, 4) of the crab population in the series IV and I of simulation.

and variations of the intraspecific competition for fewer new adults than in the series I. Juveniles and adults join in feeding on hydranths The age structure of the population includes therefore specimens of all ages and sizes. This functioning mode of the model community can be attributed to the differences in the assumed mechanisms of the trophic nudibranch - infusorian relation in the series I and IV. A change of the community functioning mode with a change of the species interaction mechanism entails a restructuring of the vital processes and the age structure of both populations. The difference in the size and age structure of the nudibranch population has an impact on the hydroid, which, in the series I, is under a stronger pressure from more numerous adults. In the series IV, the hydroids' vital processes are smoother and take more time, the biomass being approximately 1.4 higher than in the series I. By the end of the model period the nudibranch population

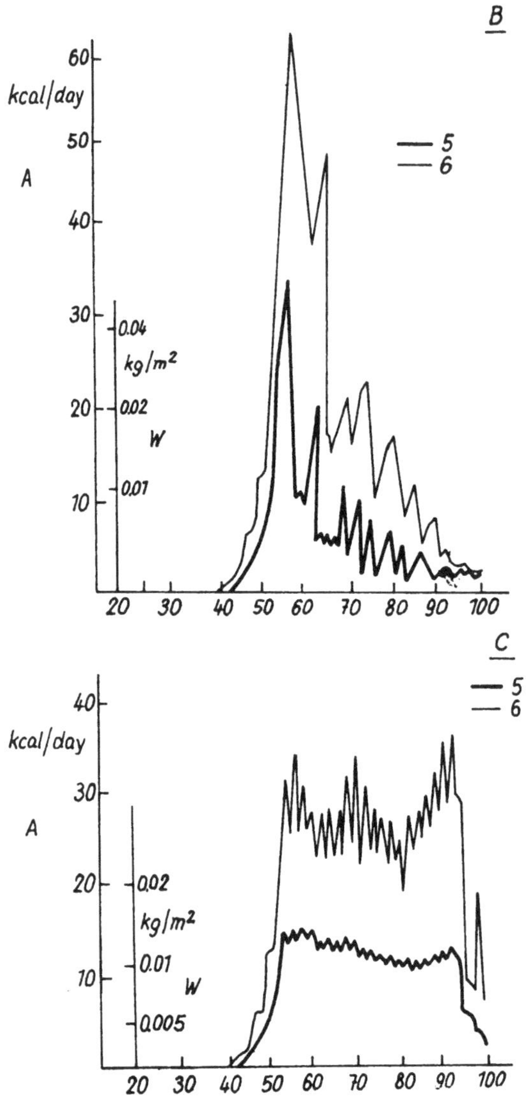

Fig. 39 B & C. Biomass (5) and assimilated food (6) of the nudibranch population for simulation series I(B) and for series IV(C).

better preserves its biomass in the series IV, whereas on T = 90 in the series I, there are but a few large specimens, in the series IV very tiny specimens are quite numerous and there is twice that number of adults. Although, in the series I, the substrate area for the infusorian is three times that of the series IV, by the end of the model time the community of the series IV is more viable.

The dynamics of the basic characteristics of the omnivorous crab population depends in both series only on the generative part of the

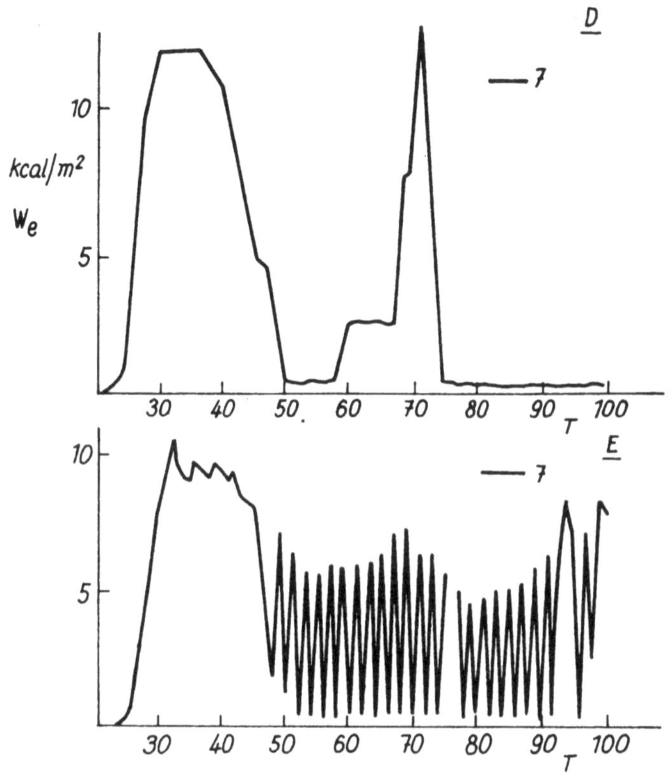

Fig. 39 D & E. Energy value (7) of the infusorian population for simulation series I (D) and series IV (E).

hydroid colony, the crab's substrate. The crab has to play a significant controlling role mainly in an inhibited community where it keeps the model population in this state in numerous runs. In periods of a normal development of dependent populations, the crab does not change the community state to an important degree, Fig. 39. The omnivorous crab does not influence the age structure of the nudibranch population in the series IV, where it feeds on these eggs. It also affects the intensity of the intraspecific competition of juvenile nudibranches. By the end of the model period, there is no crab population to speak of in the series I, whereas in the series IV its biomass stays at the level of T = 30 although the age structure does not include large specimens. In the runs of the series I and IV the balanus was inhibited all the time.

A brief discussion of the series I and II (Table 24; curves 2 and 4, Fig. 38) would be enlightening. In the series II the substrate area of the model infusorian population is expanded to cover the entire regenerative part of the hydroid. The nudibranch continuously consumes the infusorian. The infusorian regeneration period is increased to two or three days. Because of the larger substrate area the intraspecific competition among juvenile nudibranches occurs later than in the series I. The numerous adult predators inhibit the hydroid fast and reduce its biomass. Simultaneously, the infusorian is kept effectively underdeveloped. The controlling role played by the crab is quite pronounced. The viability of the nudibranch and the infusorian is undermined. On T = 70 - 80, Fig. 38, the hydroid biomass gradually restores and the hydranths regenerate. The infusorian population also gradually returns to a normal state. An intraspecific feeding regulation resumes in the hydroid, curve 7, Fig. 38. New nudibranch generations develop. At that time the model community runs out of the plankton food. Parts of the hydroid colony break away, thus depriving the other populations of the substrate.

In the series III the processes are, on the whole, smoother and the age structures are better balanced than in the series I and II. The reduced substrate areas available for the infusorian prevent the nudibranch population from dramatic expansion. This series is most favourable for the hydroid, which finds food available throughout the model period.

Consequently, changes of the mechanism of interaction between the nudibranch and the infusorian and of the substrate area for the food species often result in qualitative as well as quantitative changes of the functioning modes for the model community. This conclusion is also true of the hydroid - balanus relation. Thus, in the series V, the balanus makes a multi-layer population and in nearly every run prevents the development of the hydroid whose biomass is merely 0.04 - 0.06 kg, neither does the community develop.

The above discussion confirms the predicting power of simulation and the ability of the model to tackle managerial problems. Control of the fouling community seems to be an issue requiring further detailed discussion, but the available results suggest that there are key processes by means of which the community functioning may be modified as well as the development of the hydroid, the trophic nudibranch - hydroid, and the topic hydroid - balanus relation. The simulation results help the the scientist and engineers in deciding upon the biological control of the fouling community. The application sphere of the simulation model and the findings of the computer experiments have, however, broader implications. Since the energy flows through the community populations can be measured, this holds likewise for the energy flows through various trophic levels, and the impact made by the environment of bottom populations on these processes can be assessed.

3.7. Dynamic Forecasting of Basic Community Characteristics for Various Food Concentrations and Salinities. Model Sensitivity to Parameter Manipulation

A better insight into the development of the community could be provided by investigating the impact made by variations of the salinity S in permille and concentrations of phytoplankton α in thousand specimens/m^3 and zooplankton, β in thousand/m^3. Temperature was considered a signal factor. The simulation design included 15 runs of the system of models for each of the five simulation experiment series, Section 3.3., which differ in the conditions under which the community forms and functions. Table 25 (Section 3.3.) represents the quantitative characteristics of parameters in the runs for different simulation series. In particular, the series IV, Section 3.3., simulates realistically the situation frequently observed in the field. Fig. 40 shows a dynamic forecast of basic community characteristics for $\beta = 50$, S = 8 permille and S = 5 permille. In the following Figures the lines will connect computed points with an interval of 10 days and the word permille or o/oo will be omitted.

With S = 5 o/oo the hydroid polyp develops slower than with S = 8 o/oo and the vital processes are slower. As the development and, consequently, regeneration are gradual, there are more living hydranths until T = 60 to 70 with S = 5 o/oo. The low salinity also restricts the nudibranch population. The assimilated food of the hydroid is a more complex function of salinity variations because the nudibranch - hydroid interaction results in the triggering of compensatory mechanisms (hydranth's regeneration) and in the amount of assimilated food increasing by the discontinuation or reduction of energy flows for other physiological needs, Sections 2.1. and 3.2. Thus, from T = 60 to T = 70, the food assimilated by the hydroid with S = 5 o/oo is, on the average, 100 to 120 kcal larger than with S = 8 o/oo, with the biomasses being approximately equal. This fact is an evidence of differences in the age structure of the hydroid population in the two runs (the hydroid grows and regenerates by branching, Section 2.1.).

Because of the slow growth of new hydranths and the gradual regeneration of hydranths consumed by the nudibranch, which is not abundant, with S = 5 o/oo the hydroid population was less inhibited than with S = 8 o/oo. The nudibranch - hydroid interaction was most intensive from T = 60 to T = 80. Figure 40, A4 shows that the rate of this relation, or the energy value of consumed hydranths, is 10 to 15 kcal higher with low salinity. Adding the hydroid weight lost to the predatory nudibranch and the amount that would exist, grow, and regenerate if it were not consumed, the rate is 20 to 25 kcal higher. A smoother hydroid development with S = 5 o/oo until T = 60 to T = 65 results in a

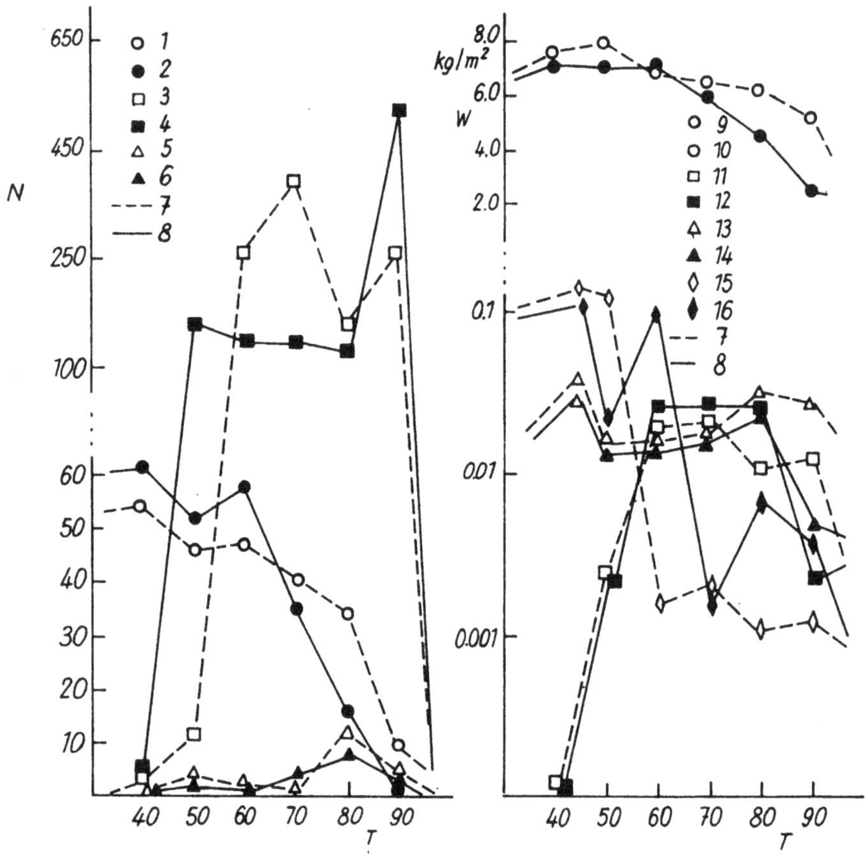

Fig. 40 - A1 & A2. Dynamic forecast of basic characteristics of model community populations with different food concentrations and abiotic factor (salinity).
The case of β = 50, A1 - A4; the case of β = 30, B1 - B4; the case of β = 10, C1 - C4.

Notation of Fig. 40

All the hatched notations are for the salinity of 5 o/oo, the remaining ones for 8 o/oo.

1, 2 - number of hydroid hydranths, millions
3, 4 - number of nudibranch specimens, thousands
5, 6 - crab population size
7, 8 - conditional connecting lines
9, 10 - hydroid population biomass
11, 12 - nudibranch population biomass
13, 14 - crab populations biomass
15, 16 - infusorian population biomass
17, 18 - food assimilated by hydroid population

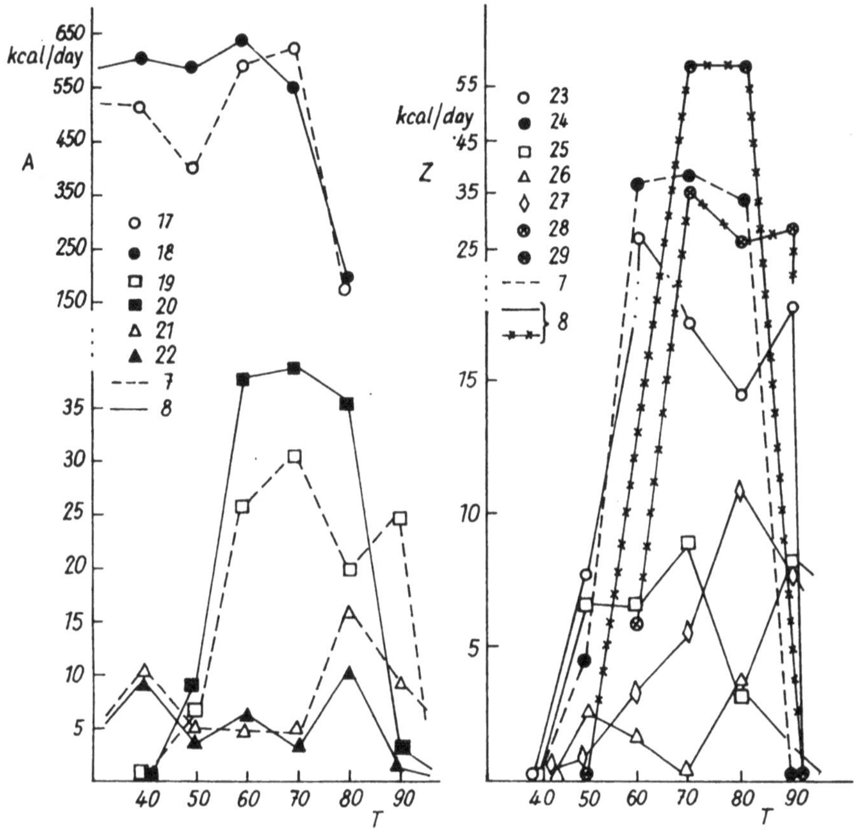

Fig. 40 - A3 & A4

(Notation of Fig. 40 - continued)

19, 20 - food assimilated by nudibranch population
21, 22 - food assimilated by crab population
23, 24 - rate of nudibranch - hydroid interaction in energy value of consumed hydranths
25 - 27 - rate of nudibranch - infusorian, crab - infusorian, and crab - nudibranch interaction, respectively, in energy value of consumed specimens of dependent species .
28, 29 - rate of nudibranch - hydroid interaction for five and eight permille salinities, respectively, in total energy loss of hydroid population (including losses due to growth, regeneration, and removal of multiplication)
30, 31 - the same as in 28, 29
32, 34, 36 - the same as in 25, 26, 27 (salinity 8 o/oo)
33, 35, 37 - the same as in 25, 26, 27 (salinity 5 o/oo)

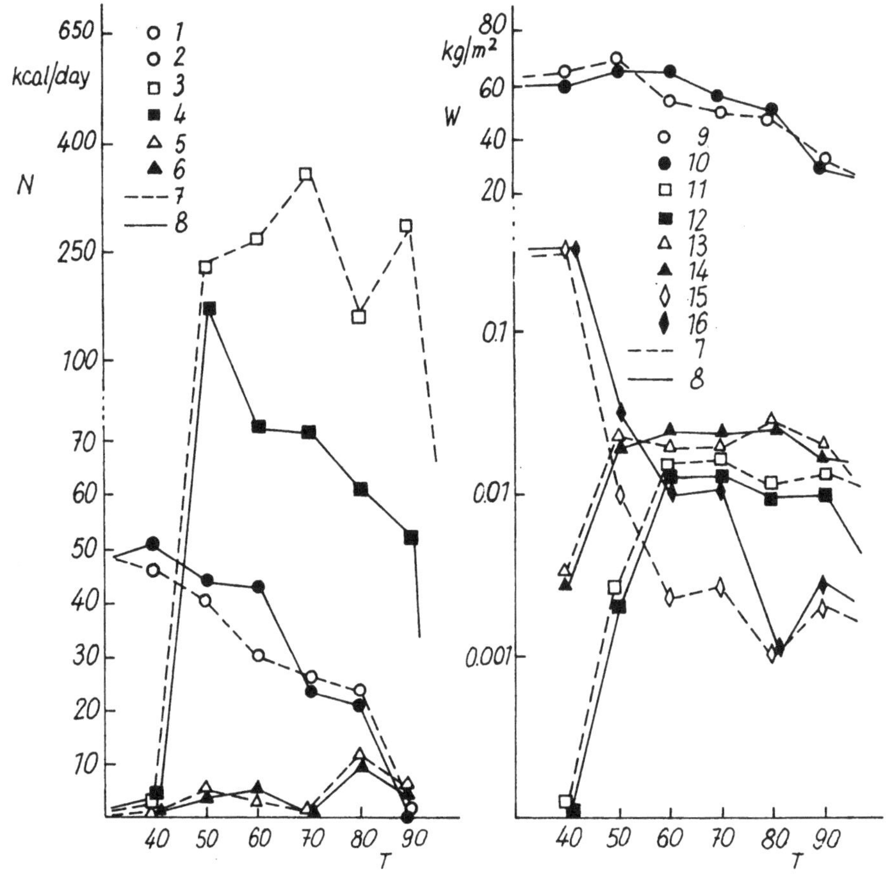

Fig. 40 - B1 & B2

balanced age structure of the nudibranch population, which maintains, with β = 50, more adults by the end of the model period. For this reason the biomass and assimilated food of this population with low salinity is higher over nearly the whole interval than with a higher salinity. The change of the nudibranch age structure due to lower salinity becomes obvious by comparing the curves of these quantities in the Figures 40 - A1 and 40 - A3 of this population. Its slower regeneration protects the nudibranch from loss of most juveniles due to shortage of food and in this way the age structure is kept balanced. For the same reason the biomass of the infusorian community is higher with S = 5 o/oo than with S = 8 o/oo, Fig. 40 - A2.

The changed salinity with reduced model food concentration for the hydroid slows down practically all vital processes in nearly all populations that are topically and trophically dependent on the hydroid (Fig. 40 - B and C). The rate of the nudibranch - hydroid interaction

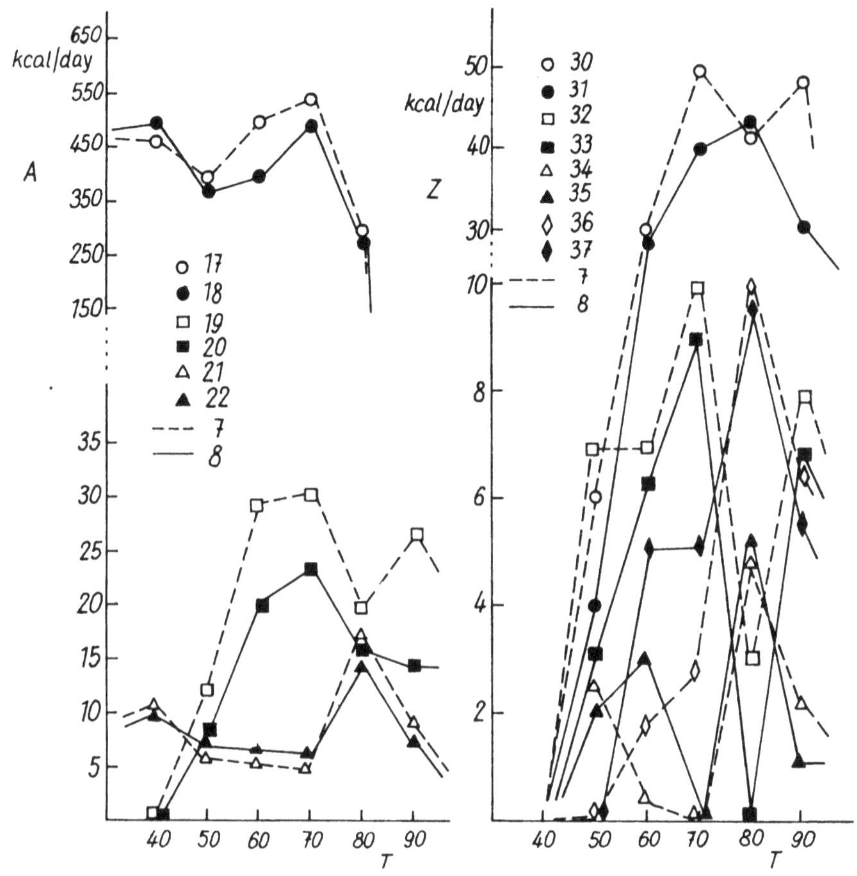

Fig. 40 - B3 & B4

with β = 30 remains, on the average, similar to that with β = 50 over the entire model period. With β = 10 all processes slow down, the interspecific relations are short-living, and the dependent populations are severely inhibited. With the salinity reduction and this low   the model community cannot develop.

The change of β from 50 to 30 changes the biomass and the assimilated food of the topically dependent infusorian and crab population within 20 to 25 per cent, while during a long time the variation of the food concentration has an even less remarkable impact, because in this series the populations topically dependent on the hydroid develop only on some of the possibly available area, Section 3.3. With β = 10, when the hydroid has not grown, all topically dependent populations are also underdeveloped. The rate of the trophic crab - and nudibranch - infusorian relations do not change much as β falls from 50 to 30, they are maintained at a level of six to ten kcal/day, and are practically insignificant for β = 10, although their biomasses for numerous days are

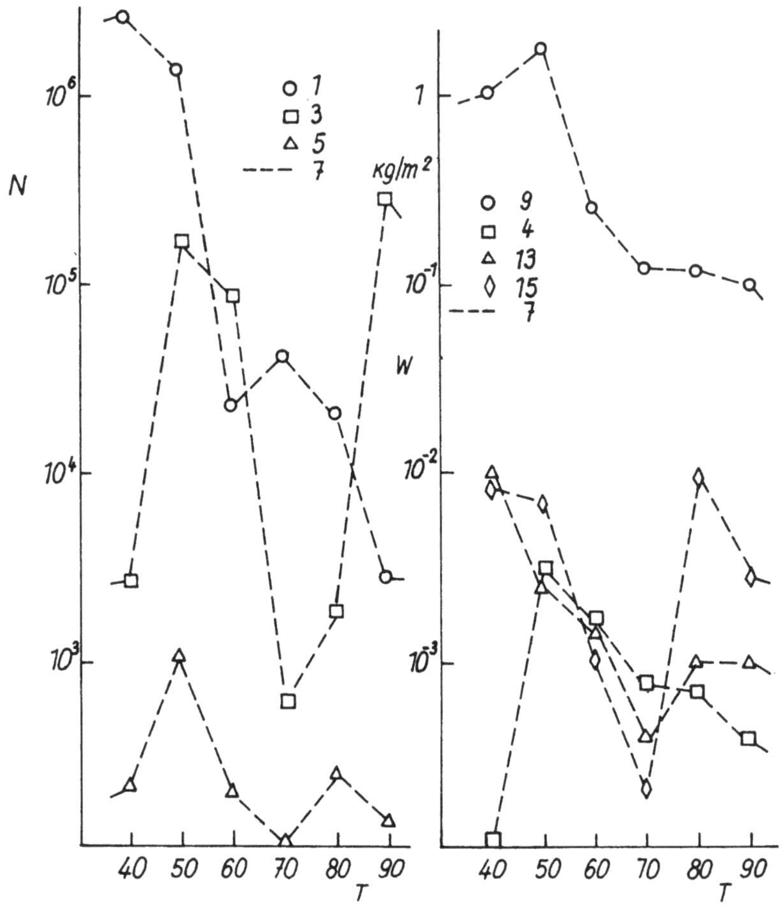

Fig. 40 - C1 & C2

comparable with those with β = 50 and β = 30. The salinity reduction results in small variations in the crab's population size. The population keeps the development of the nudibranch and infusorian in check in those runs where the community develops slowly. With β = 30 and S = 5 the infusorian is on the average in favourable conditions because the hydroid biomass changes smoothly and the nudibranch is not abundant. By the end of the model period the hydroid population is in different states in the runs where the values of β are different. While in most runs with high β the final biomass was two to 2.5 kg/m^2, with β = 10 that value is quite different and is as low as 110 g/m^2. The final biomass of the omnivorous crab, which populates the generative part of the hydroid colony in most series of the simulation experiment, Section 3.3., (Gal'perin, 1983) is directly dependent on the biomass of this part of the hydroid population. At the end of the model period the biomass of other species, the nudibranch and the infusorian, is a function

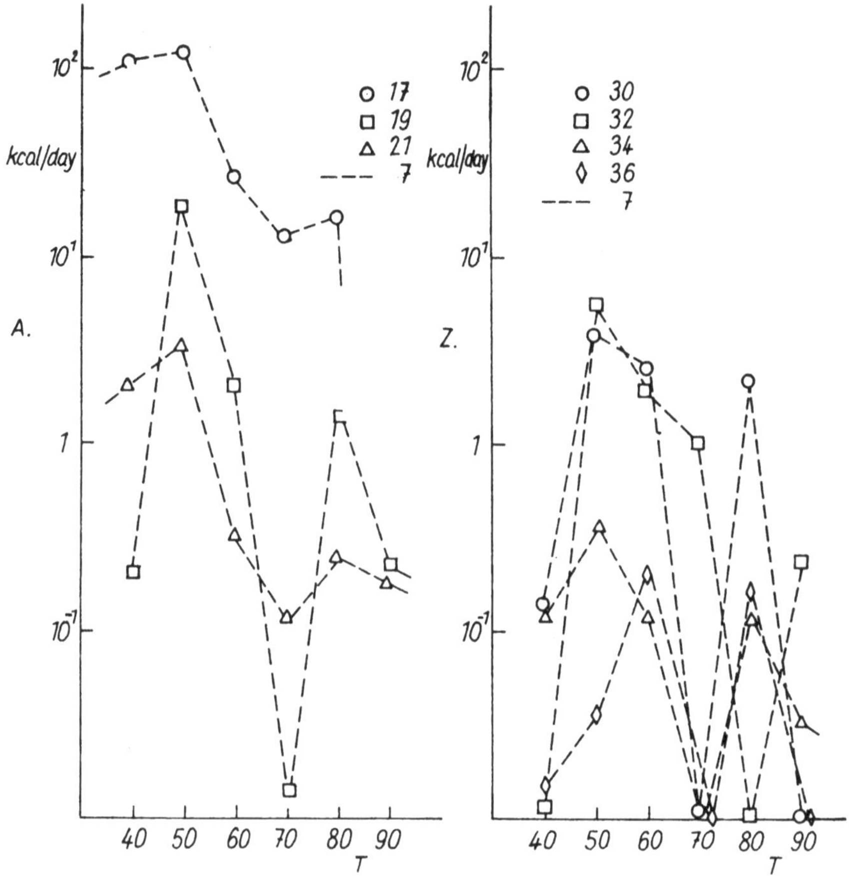

Fig. 40 - C3 & C4

of the intra-population mechanisms more than of the substrate area. Let us now have a closer look at the population interactions as a function of the salinity and the hydroid food concentration.

The peak of the intraspecific competition in the nudibranch population, Fig. 41, starts earlier in the period of the nudibranch - hydroid relation, when the food concentration is lower. This competition among adult nudibranches is the mildest with $\beta = 30$, because in this run the hydroid population develops smoothly and the nudibranch age structure is balanced. Although all the runs are comparable in terms of the energy value of dying nudibranches on the days when the rates are maximum, the intraspecific competition with $\beta = 30$ is still 10 - 15 kcal/day lower. Even in the general pattern, the dynamic curve of the intraspecific competition for $\beta = 10$ is obviously different from other runs, Fig. 41. The nudibranch population dies fast because the infusorian is in short and practically non-renewable supply on the substrate. As noted above, the case of $\beta = 30$ is most favourable for the

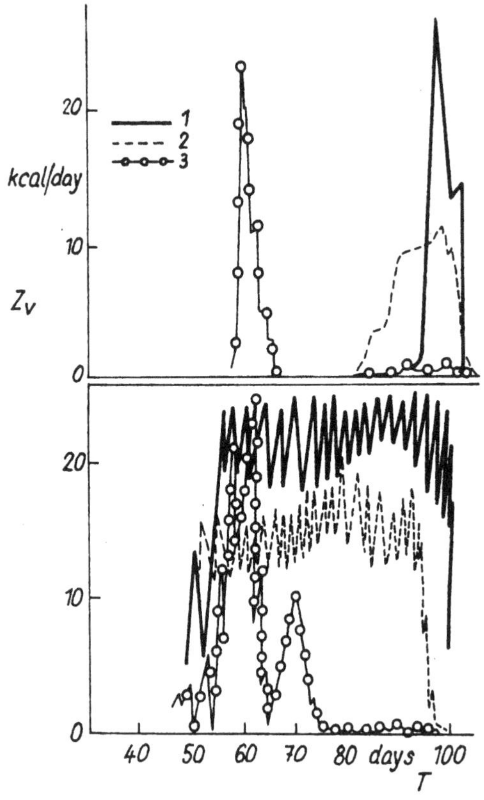

Fig. 41 Dynamic forecast of intraspecific competition rates in nudibranch population with various hydroid food concentrations of dependent hydroid.
1 - intraspecific competition of adult (top) nudibranches on the hydroid and of juveniles (bottom) on infusorian for β = 50, salinity 8 o/oo, series IV; 2 - the same, but β = 30; 3 - the same, but β = 10
The intraspecific competition rate is expressed as the energy value of the biomass of dying specimens.

nudibranch as far as the number and energy value of the dying specimens are concerned. In this case, the intraspecific competition of the juveniles averages from 11 - 12 to 16 - 18 kcal/day while for β = 50 this range is 18 - 28 kcal/day.

When the salinity changed to five permille in case of β = 10, the populations did not noticeably interact and both were inhibited. In case of β = 30, this change resulted in extending the period of the intraspecific competitions dynamics for both the juveniles and adults by a factor of 1.5, although the total losses of the nudibranch population were similar in both runs. With β = 50, the change of salinity from eight to five permille reduced the absolute value of the intra-

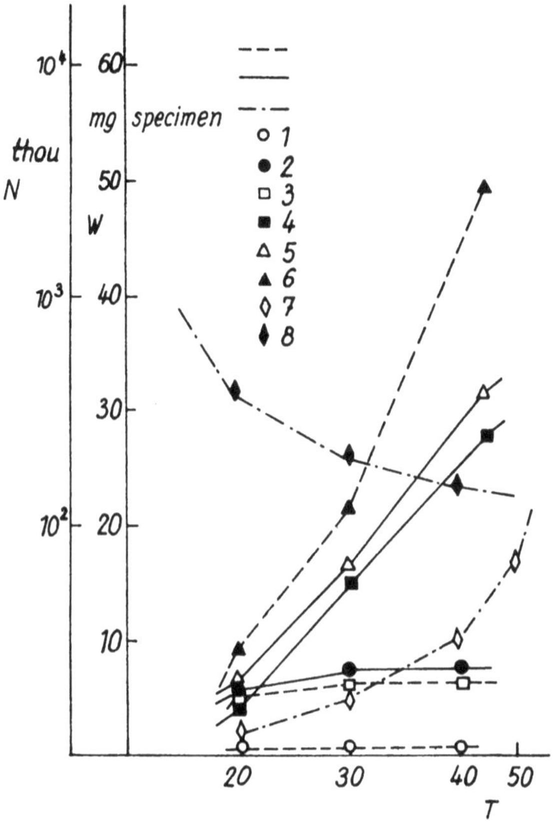

Fig. 42 Dynamic forecast of the biomass of a balanus specimen in a single layer and the same settlement and multi-layer and age settlement.
1, 2 - biomass, with $\alpha = 120$, $\beta = 50$, salinity 8 and 5 o/oo, respectively. 3, 4 the same as 1, 2 but for $\alpha = 120$, $\beta = 30$. 5 - the same with $\alpha = 170$, $\beta = 20$ and salinity 5 o/oo. 6 - the same with $\alpha = 120$, $\beta = 10$, and salinity 8 o/oo. 7, 8 - average specimen biomass in a series simulating a multi-layer balanus settlement and population size with salinity 8 o/oo.

specific competition in the nudibranch population although the relative value was comparable with that if S = 8.

The impact of variations in the balanus' food concentration, an important variable for the juveniles, is shown in Fig. 42, which shows a dynamic forecast of basic characteristics of this population as a function of the phytoplankton concentration α and the development of the hydroid interacting with the balanus, Sections 2.2. and 3.2. For the case of a multi-layer balanus settlement which inhibits the hydroid, Fig. 42 shows the size of the population along with the biomass of one average specimen. In case of a single-layer settlement, when the hydroid retards or completely suppresses the balanus, salinity becomes a major factor. This fact is most obvious for $\alpha = 120$ and $\beta = 30$, if with S = 8 the balanus is practically suppressed, while with S = 5, the hydroid being inhibited, the barnacle weighs 28 mg against as little as 6.2 mg with S = 8. These results obtained from simulating the balanus - hydroid interaction confirm the biological theory on the direct dependence of the hydroid and balanus development and, consequently, of the processes which dictate their development. When the hydroid is inhibited by a multi-layer settlement, the maximum biomass of the dependent population with S = 8, $\alpha = 120$, and $\beta = 100$ was 1.63 kg/m^2, whereas with its normal development and with these food concentrations, the hydroid biomass was as high as 8 - 8.5 kg/m^2. Changing the salinity improved the environment for the balanus.

Consequently, this analysis suggests that the simulation study (a system of models) is satisfactorily sensitive to variations of basic parameters, the criterion being the fit of simulation results and field observations when the environmental conditions specified in the models parametrically change.

Chapter 4. BIOLOGICAL CONTROL OF BOTTOM COMMUNITY DEVELOPMENT AND PRINCIPLES OF THE EVENT APPROACH TO SIMULATION IN MARINE ECOLOGY

4.1. On Two Mathematical Approaches to Controlling Biologically the Evolution of Bottom Communities

Introduction

Ecosystem research methods remain unreliable despite the increasing number of publications. Indeed, even relatively simple ecosystems build up numerous hierarchical levels. In the mathematical formalization of control even for low levels, the mechanisms of feedbacks, self-adaptation, and self-organization should not be overlooked because, in response to exogenous signals, they restructure the ecosystem (Straskraba, 1979). If an ecosystem is considered a self-adapting process, studies of the control potential should cover, in particular, the channels through which the intrapopulation trophic patterns, the inter-population relations, and behavioural responses can be influenced.

A major issue is the choice of the objective function which would give an insight into the goal, if this term is legitimate, of the ecosystem. Even the destruction of the ecosystem is biological, or for the ecosystem itself, the optimum strategy. Thus man usurps, in a sense, the right to formulate the goal of the living system. This issue may be swept aside by the researcher, but even in most general terms, it remains a difficult task.

The Conventional Control Problem

A continuous dynamic system

$$\dot{x}^i = f^i\,(t,\ x^i,\ u) \qquad (i = 1,\ \ldots,\ N)\ , \tag{4.1.1.}$$

operates under boundary conditions

$$x^i(0) = x_0^i;\quad u \in V^{tx};\quad x \in V_x^t\ , \tag{4.1.2.}$$

and implements a functional

$$I = \int_0^T f^0(t,\ x)\ dt + F(x(T)) \xrightarrow[\min]{\max}\ . \tag{4.1.3.}$$

It is required to move the system from time t = 0 to time t = T with the largest (least) possible value of the functional (Krotov and Gurman, 1973). If several biological indices have to be minimized in an ecosystem, the functional takes the form

$$I = \int_0^T \alpha \cdot \left[f(x(t)) \right] dt + \bar{\alpha} \cdot F(x(T)) \xrightarrow[\min]{\max} , \qquad (4.1.4.)$$

where $\alpha \cdot \left[f(x(t)) \right] = \sum_{j=1}^{m} \bar{\alpha}_j \cdot f^j(x_j(t))$; α_j is a vector of Boolean variables; and $F(x_j(T))$ is a term which specifies the condition under which the biological indices may reach the desired domain at time T (in a limit case i = N).

Reformulated in more specific terms, the problem (4.1.1.) - (4.1.4.) becomes more involved. Let $x^i(t)$ be the indicator of the i-th population (i = 1, 2, ..., m). The function $f^i(\beta^i \cdot x^i(t), t)$ characterizes the natural evolution, β^i is a Boolean variable for the i-th population. Let us introduce two sets of control functions for migration processes (natural and removing individuals from a population)

$$\varkappa^i(u^i(x^i(t), t), t)$$

and for abiotic factors

$$\eta^i(u^l(x^i(t), t), t) , \quad l = m + 1, m + 2, \ldots, n.$$

Then the vector of controls for the i-th population is

$$\tilde{\tilde{u}}^j = \alpha_1^j \cdot \varkappa^i + \alpha_2^j \cdot \eta^i , \qquad (4.1.5.)$$

where α is a vector of Boolean variables which recognizes the time dependence of the structure of the control action. Now introduce functions of trophic relations between the j-th and the i-th populations. In a general form this is (i is the index of the dependent population)

$$P_j^i (x^j(t), t, \tilde{\tilde{u}}^j(\varkappa^i, \eta^i, j), \sigma_j^i (R, t)) , \qquad (4.1.6.)$$

and, taking into account topical relations,

$$S_j^i (x^j(t), t, \tilde{\tilde{u}}^j(\varkappa^i, \eta^i, j), \delta_j^i (R, t)) , \qquad (4.1.7.)$$

where σ_j^i and δ_j^i are characteristics of territorial and spatial (structural) relations between the components, and R is the metric of the space E_n. The process equation is now

$$\dot{x}^i = f^i(\beta^i, x^i, t) + \varkappa^i(u^i, t) + \eta^i(u^l, t)$$
$$+ \sum_{j=1}^{m} \left[P_j^i (x^j, t, \tilde{\tilde{u}}^j, \sigma_j^i) + S_j^i (x^j, t, \tilde{\tilde{u}}^j, \delta_j^i) \right] . \qquad (4.1.8.)$$

The boundary conditions can be represented in a general form as

$$V_x^t: x^j(t) \geqslant 0,\ x^j(0) = x_0^j,\ x^i(t) \geqslant 0,\ x^i(0) = x_0^i\ ;$$

$$V_i^{tx}: \quad 0 \leqslant u^j(x^i, t) \leqslant x^i(t), \qquad (4.1.9.)$$

$$\xi^l(a_1, t) \leqslant u^l(x^j, t) \leqslant \varphi^l(a_2, t),$$

$$(\text{with } \varkappa,\ \eta \in V^{tx}),$$

where a_1^l and a_2^l are environment factors.

Trophic relations and the endogenous evolution of populations would make equation (4.1.8.) even more complicated. The introduction of several feedbacks would make the functional very nonlinear. Even without these complications the very first steps in solving the problem (4.1.6.) - (4.1.9.) encounter numerous, often unsurmountable, technical difficulties. First, the proof of existence and uniqueness and attempts to design a control result in a system of partial differential equations. Even if the right-hand sides of equation (4.1.8.) are not very complicated, such a system does not ensure convergence in the numerical implementation, or even of analytical solutions. Furthermore, the form of the terms of equation (4.1.8.) and their parameters are essentially very difficult to formulate for a specific ecological unit. Finally, a numerical solution, if available, is not necessarily asymptotically stable. This reasoning seems sufficient to show that the approach of the equations (4.1.6.) - (4.1.8.) is not promising.

A General Outline of the Proposed Approach

Real ecosystems can rather be tackled in a different sequence

a) development of a detailed ecological model of the ecosystem incorporating most cause-and-effect relations;

b) mathematical formalization of the processes by a logical-heuristic method; Gal'perin, 1983 a-e);

c) scenario simulation in the entire parameter ranges (e.g. Chapters 2 and 3);

d) setting up a bank of simulation results to store the data on the effect of various intra- and inter-population mechanisms on the system dynamics as well as integral ecosystem characteristics;

e) application of one of the methods to determine the optimal effect of a factor on the ecosystem performance;

f) ecological analysis of the results and formulation of the control sequence.

Every step of this algorithm may be thoroughly described for a specific example.

The Ecological Dimension

Most ecological models study variations with time. Consequently, time is a variable in optimization and control problems (Gurman, 1977, 1979). This approach is however not necessarily legitimate in ecology, in particular, water ecology (Gal'perin, 1985b). The evolution in time and the associated changes of state make a subset of system states where time is merely a scale in which the events are ranked. Therefore, the proposed approach can be tested in a fairly general case of an ecosystem, where the effect of time on a model interval is either insignificant or is just one of the factors that compel the system to change the state.

Bottom coenoses are typical of such ecosystems. They do not change much with time, respond slowly to exogenous impacts, and the topic biocoenotic repletions, which are of parameter importance, are regulated chiefly by space variables. Shorygin (1951) and Beklemishev (1951) were pioneers in the qualitative identification of various kinds of symphysiological relations. The significance of inter-species relations was estimated for a simple case of bottom biocoenoses, namely the fouling community, which can be easily observed and whose economic effect has been increasing with man's penetration of the shelf. The fouling has been removed mechanically or chemically or stained, but these techniques are costly and result in an ecological misbalance the consequences of which may be unpredictable.

As mentioned above, qualitative estimates of biological defouling were discussed, in particular, by Turpayeva (1972). The data collected over two decades were processed and the fouling of water supply installations in iron and steel works in the Bay of Taganrog of the Sea of Azov was described in detail. These data suggested a model concept; a mathematically formalized fouling biocoenosis model has been developed and programmed (Gal'perin, 1983 a-e).

A Specific Ecological Object

Recall that a model of interspecific relations (cf. Chapter 3, Table 21) consists of the following elements (Gal'perin, 1983 b): Direct topic "hydroid - infusorian" relation (No. 1): substrate supply to most of the species. Direct topic "hydroid - nudibranch" relation (No. 2): the hydroid is the substrate for the nudibranch population to feed on and expand. Direct trophic "nudibranch - hydroid" relation (No. 3): food supply to mature nudibranches. The hydroid hydranths are consumed and this triggers the regeneration mechanisms in the suppressed population, a change in the ratio of the energy balance components, the partial or complete cessation of hydroid growth, and the tearing off of the damaged nonregenerated parts of the colony. Direct trophic relation "nudibranch -

infusorian" (No. 4): juvenile nudibranches are fed and the number of infusoria is reduced. Direct topical "hydroid - crab" relation (No. 5): the entire crab population is fed. Direct trophic "crab - nudibranch" relation (No. 6): adult crabs feed on nudibranch ova deposits, thus reducing the reproducibility of the population and the number of nudibranches. Direct trophic "crab - infusorian" relation (No. 7): juvenile crabs are fed. Indirect trophic "crab - hydroid" relation (No. 8): No. 3 relation is weakened because the crabs consume some of the nudibranch population. Indirect trophic adult "crab - infusorian" relation (No. 9): the crab population curbs the reproduction of the nudibranch population and reduces the subsequent pressure of the juvenile nudibranches on the infusorian. Indirect topic "nudibranch - infusorian" relation (No. 10): nudibranches cover the unconsumed areas of the hydroid with mucus and thus prevent the development of infusorian there. Intraspecific relations play a major role in various modelled intervals of the life activity. This description disregards the interspecific relations of the balanus because, if it creates a single-stratum settlement, the balanus has an insignificant effect on the community and this is precisely the case for which the numerical data have been reported in this Chapter.

Turpayeva (1972) discussed the qualitative ecological control of the fouling development and proposed three possible ways to reduce the biocoenosis development: changing the water supply in the area in order to influence the dominating hydroid species by abiotic factors; hindering the hydroid expansion by reducing the food (zooplankton) concentration by the construction of draining installations; and biological control proper by introducing a predator species. The quantitative aspects were omitted in that the most significant sets of interspecific relations in terms of weight or energy units were not determined. Concerning the introduction of a predator, the duration of this action or the numbers of the predators for the desired effects were not specified. To obtain these figures, a simulation model of the fouling community has to be developed.

Processing the Experimental Results

The control channels of the fouling biocoenosis were classified into two groups, those which tackle the natural substrate and those which handle the interspecific relations in the community. The numerical results were stored in different files. Each file was an image of a passive factor experiment because the data were arranged for different levels of factors and time intervals in the experiment, time proper not being used as one of them. All population characteristics were calculated for one square meter of the artificial substrate, and energy equivalents of weight indices were obtained.

The choice of the method for a quantitative evaluation of biological control channels was dictated by the following ecological facts:

- depending on the time period and the range, the effect of any factor may be of either sign;
- there are boundaries between the significant and insignificant effect of the factors that vary with the combination of environmental conditions;
- the biocoenosis evolution is often most profoundly influenced by a set of relations between the dominating forms and, very rarely, by single relations.

The high complexity and hierarchical nature of the simulation model made it possible to evaluate the total effect of any component on the dynamics of basic indicators. On the other hand, the mechanisms behind the action of these components, or factors, on the system indicators such as the biomass, population size, and energy flux remain unknown. Therefore, the method of multi-dimensional statistical analysis (Liepa, 1981) seemed promising. The multiple determination ratio R^2 (e.g. Afifi and Eisen, 1979) characterizes the total effect of the factors on an indicator.

Studies of Complex Systems by the Dynamic Methods of Specific Impact

In the proposed model, the system is represented as a totality of factors. The mathematical simulation tools are correlation and regression analysis, where the coefficients of pairwise and multiple correlation are computed and a linear regression is obtained. The data for a specific model experiment are represented as a matrix whose columns stand for factors describing the system and the rows for the experiments (values of these factors).

The factor Y, for which the regression is obtained, is referred to as the system response. (Responses are integral indicators but theoretically any factor can be a response. All the factors describing the system can be successively analyzed.)

The coefficients of the linear regression equation and the covariances help to "redistribute" the multiple correlation coefficient among the factors; in other words, the "impact" of every factor to response variations is determined. The specific impact of the factor is the fraction to which a response depending on a totality of factors in the model changes due to this factor.

The impact of a specific factor is described by a specific impact indicator (Liepa, 1981), which is computed by the following algorithm:

$$\gamma_j = \alpha \cdot \left[b_j \cdot c_{oj} \right], \quad j = 1, 2, \ldots, k \quad , \qquad (4.1.10.)$$

where γ_j is the specific impact indicator of the j-th factor; k is the number of factors studied simultaneously; b_j is the j-th multiple regression coefficient which is computed by the formula

$$X_0 = a + \sum_j b_j \cdot X_j \quad , \qquad (4.1.11.)$$

where X_0 is the system response to be investigated, a is a free term of the regression, and X_j is the value of the j-th factor. The coefficient α of equation (4.1.10.) is computed by the formula

$$\alpha = R^2 / \left[\sum_j \left| b_j \cdot c_{oj} \right| \right] , \qquad (4.1.12.)$$

where R is the coefficient of multiple correlation computed by the formula

$$R = \left[\frac{n^2 \cdot \sum_j b_j \cdot c_{oj}}{n \cdot \sum_j x_{oj}^2 - \left(\sum_j x_{oi} \right)^2} \right]^{1/2} , \qquad (4.1.13.)$$

where n is the number of observations, which cannot be below 2 x K; x_{oi} is the value of the system response in the i-th observation, c_{oj} is the covariance coefficient of the system response indicator and the j-th factor. It is given by the relation

$$c_{oj} = \left(n \cdot \sum_i x_{oi} \cdot x_{ji} - \sum_i x_{oi} \cdot \sum_i x_{ji}\right) / n^2 \quad . \qquad (4.1.14.)$$

The specific contribution indicator is obtained mainly from the coefficient of multiple determination, which is computed by the formula

$$R^2 = \left(\sum_j b_j \cdot c_{oj} \right) / D^2 \qquad (4.1.15.)$$

where D^2 is the response variance. The specific impact of the j-th factor on the determination coefficient depends only on the ratio of addends in this formula. This implies that the addend whose magnitude is the largest is associated with the largest specific impact. Since the regression coefficients may have different signs, their magnitudes have to be taken in the totals. For this reason the coefficients γ of the specific impact are bound to be positive. However, it is important that the direction in which the factor acts by the computed γ is dictated by the sign of the regression coefficient. If this sign is positive, the impact on the response variable is positive and if it is not, the increase of the factor results in a reduction of the response function. The influence of the background factors, which are not represented in the data, is computed by the formula

$$\tilde{\gamma}_i = 1 - \sum_j \gamma_j \quad . \tag{4.1.16}$$

The importance of γ is determined from the relation for the empirical value of the Fisher criterion

$$F_j = \frac{\gamma_j \cdot (n - k - 1)}{1 - \sum_j \gamma_j} \quad . \tag{4.1.17}$$

A rearrangement, in a special way, of the initial data matrix at every experimental step makes it possible to investigate successively the dynamics of the significance of the impact the factors have on all system indicators that become responses successively. This method increases the statistical significance of the results obtained from the special algorithm for the recomputation of the initial data matrix. The algorithm embodies serial repeatability of the experiments by fixing the factors at certain levels. If the experiment is passive, the rows of the initial matrix are chosen in a special way so that, in every computation, rows with the closest values of factors (indicators) influencing the response are grouped together. The dynamics of specific contributions is computed by using the principle of data elimination.

In the proposed way, the computation of the dynamics of the insignificant information is gradually eliminated. The value of γ does not change remarkably until the significant information is rejected. A dramatic reduction of γ is associated with a threshold with which this elimination of useful information occurs. The algorithm of this operation is an iterative γ recomputation by formula (4.1.10) and a rejection of information exceeding the threshold computed. In the algorithm, the significance of the result and of the information eliminated is increased by recomputing the initial data matrix into a series-averaged matrix, the series being, for instance, the totality of matrix rows grouped around the closest values of the factor in the case of a passive factorial experiment. The series may also consist of repeated changes of the indicator with the others fixed at a specified level. Because in further discussion the series-averaged matrix is processed in order to obtain final results, the compilation of series from the data of a field experiment is a major task for the user because, both, the numerical and meaningful (qualitative) result of the computation may be influenced. With increasing threshold the amount of rejected information also increases, therefore one has to check whether the amount of information in the series-averaged matrix is sufficient, see below.

Consequently, the information on the factor considered in this version of the method is rejected by the formula

$$X_{1i} = \left[\sum_p X_{1ip} - m \cdot h \right] / n_i , \quad \begin{array}{l} p = 1, 2, \ldots, m; \\ i = 1, 2, \ldots, N, \end{array} \qquad (4.1.18)$$

where X_{1i} is the value of the i-th series in which the factor X_1 is observed and for which the critical (rejection) threshold is determined after the elimination of data with a threshold of h; n_i is the number of observations in the i-th series; m is the number of values of the factor X_1 which exceed h ($0 \leqslant m \leqslant n_i$); N is the number of observation series (rows of the N x (K + 1) matrix of the initial information, where K is the number of factors investigated simultaneously).

The standard statistical quantities and their significances are determined in the conventional way.

Quantitative Evaluation of the Control Channels

Each column in the (750 x 12) simulation result matrix represented the dynamics of one of the biocoenotic indicators. All columns were successively studied in the regression model as the system response and the remaining columns were the factors at those moments. The indicators were the population biomasses and the basic sets of trophic relations. The topic relations were evaluated from the population biomasses, while the sets of direct and indirect trophic relations in a community are given in terms of the intensities of the basic trophic relations Nos 3, 4, 6, and 7 (formula (4.1.10.)). Consequently the regressors of the statistical model came in succession:*) the hydroid weight (G), the weight of juvenile nudibranches (TJ), the weight of mature nudibranches (T), the weight of juvenile crabs (KJ), the weight of adult crabs (K), the weight of infusoria (S), the sets of trophic "nudibranch - hydroid" (SV3), "nudibranch - infusorian" (SV4), "crab - nudibranch" (SV6), and "crab - infusorian" (SV7) relations, salinity (S4), and the zooplankton (hydroid feed) concentration (β). The statistical model thus dealt with integrated biocoenosis indicators and so the results were averaged over various environmental conditions. The correlation matrix is shown in Table 26. The computation of Student criteria for the matrix showed that most coefficients had a significance of 99.5 per cent.

Fig. 43 shows the effect of γ_j, formula (4.1.10), on the hydroid, the dominant species in a fouling community, in particular, the dynamics of γ_j with various h_j^i for every factor (see the above algorithm for eliminating and computing the dynamics and Table 27 of h_j^i, where

---

*) Let us change the notation of basic quantities in order to distinguish the dynamic system variables of the model which simulates the fouling community from averaged statistical estimates.

Table 26 Correlation matrix for the original matrix of impact through conditioning species

G	1.00	0.61	0.58	0.62	0.76	0.61	0.49	0.64	0.48	0.73	-0.1	0.44
TJ	0.61	1.00	0.86	0.71	0.35	0.31	0.85	0.84	0.74	0.66	-0.05	0.15
T	0.58	0.86	1.00	0.60	0.30	0.35	0.91	0.66	0.61	0.62	-0.16	0.23
KJ	0.62	0.71	0.60	1.00	0.45	0.23	0.62	0.70	0.82	0.87	0.01	0.00
K	0.76	0.35	0.30	0.45	1.00	0.50	0.31	0.39	0.42	0.44	-0.02	0.23
S	0.61	0.33	0.35	0.23	0.50	1.00	0.33	0.12	0.13	0.32	-0.15	0.22
SV3	0.49	0.85	0.91	0.62	0.31	0.33	1.00	0.64	0.70	0.55	-0.15	-0.13
SV4	0.64	0.84	0.66	0.70	0.39	0.12	0.64	1.00	0.72	0.65	0.05	0.16
SV6	0.48	0.74	0.61	0.82	0.42	0.13	0.70	0.72	1.00	0.68	-0.01	-0.00
SV7	0.73	0.66	0.62	0.87	0.44	0.32	0.59	0.65	0.68	1.00	-0.04	0.14
SL	-0.1	-0.05	-0.16	0.01	-0.02	-0.15	-0.15	-0.05	-0.01	-0.04	1.00	-0.06
β	0.44	0.15	0.23	0.00	0.23	0.22	0.13	0.16	-0.00	0.14	-0.06	1.00

Table 27 Threshold values h_j^i of factors in computing the dynamics of coefficients γ_j^i

	1	2	3	4	5	6	7	8
G (kg)	0.000	0.987	1.973	2.959	3.946	4.933	5.919	6.906
TJ (kg)	0.000	0.002	0.003	0.005	0.007	0.009	0.010	0.012
T (kg)	0.000	0.002	0.003	0.005	0.007	0.008	0.010	0.012
KJ (kg)	0.000	0.002	0.005	0.007	0.009	0.012	0.014	0.016
K (kg)	0.000	0.008	0.015	0.023	0.031	0.038	0.046	0.054
S (kg)	0.000	0.025	0.049	0.073	0.097	1.122	0.146	0.171
SV3 (kg)	0.000	0.019	0.037	0.056	0.074	0.093	0.111	0.129
SV4 (kg)	0.000	0.019	0.038	0.057	0.076	0.096	0.114	0.133
SV6 (kg)	0.000	0.001	0.001	0.001	0.002	0.003	0.003	0.004
SV7 (kg)	0.000	0.008	0.017	0.025	0.033	0.042	0.050	0.058
SL permille	5.000	5.375	5.750	6.125	6.500	6.875	7.250	7.625
β (thousand/ m^3 water)	1.000	25.000	50.000	75.000	100.000	125.000	150.000	175.000

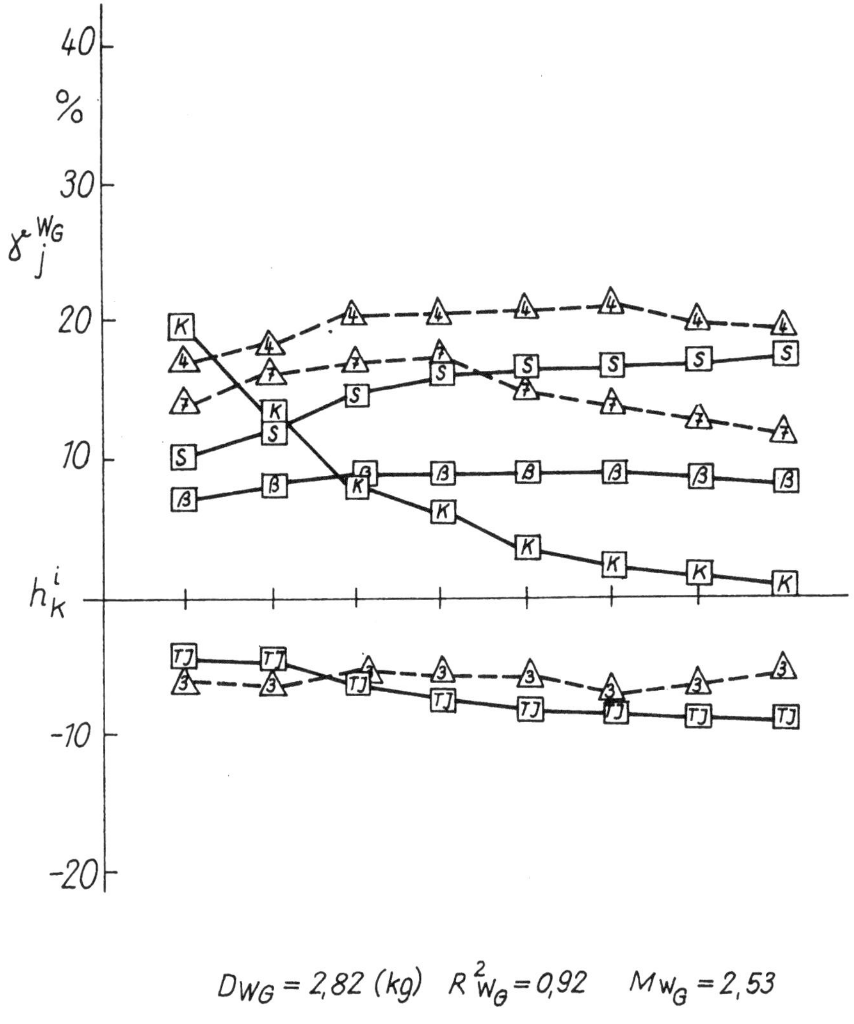

Fig. 43 The impact of threshold values of h_j^i, Table 27, of population biomasses and rates of interspecific relations on the mean (M_j) and variance (D_j) of biomasses of other species in the community. The impact is determined by γ_j, formula (4.1.10). For detailed explanation see text. Figures 44 - 48. The notation is as in Fig. 43.

i = 1, 2, ..., 8 , or a total of eight tresholds were used). The analysis of statistical data, Fig. 43, reveals that the crab makes the strongest impact on the hydroid population. These two species are related by a positive dependence. The trophic set SV7 has a marked effect but only as long as the "crab - infusorian" relation is strong. The

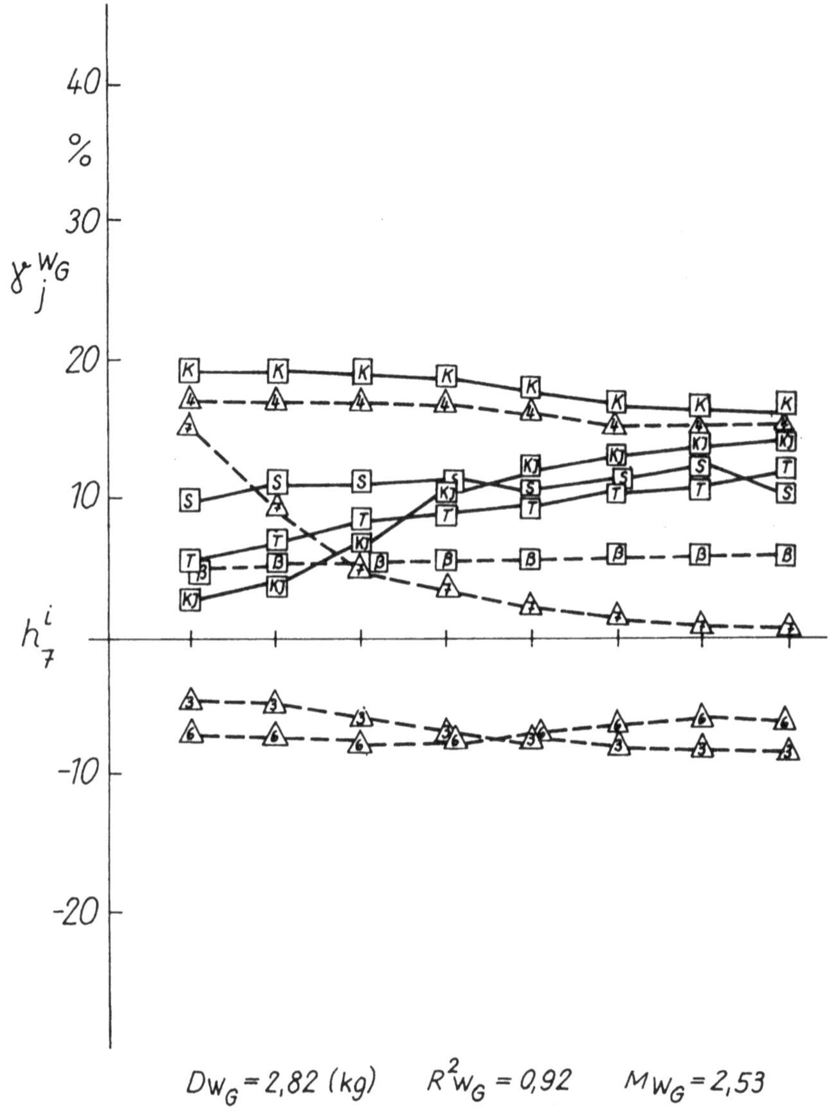

Fig. 43 (continued)

positive effect of the "nudibranches - infusoria" set on the hydroid population is due to an indirect impact of the crab on the competition among juvenile nudibranches which increases with the strength of these relations, and to the regeneration mechanism in infusoria when the prèdator's pressure increases. The negative mean dependence between the expansion of the hydroid population and the trophic "crab - nudibranch'

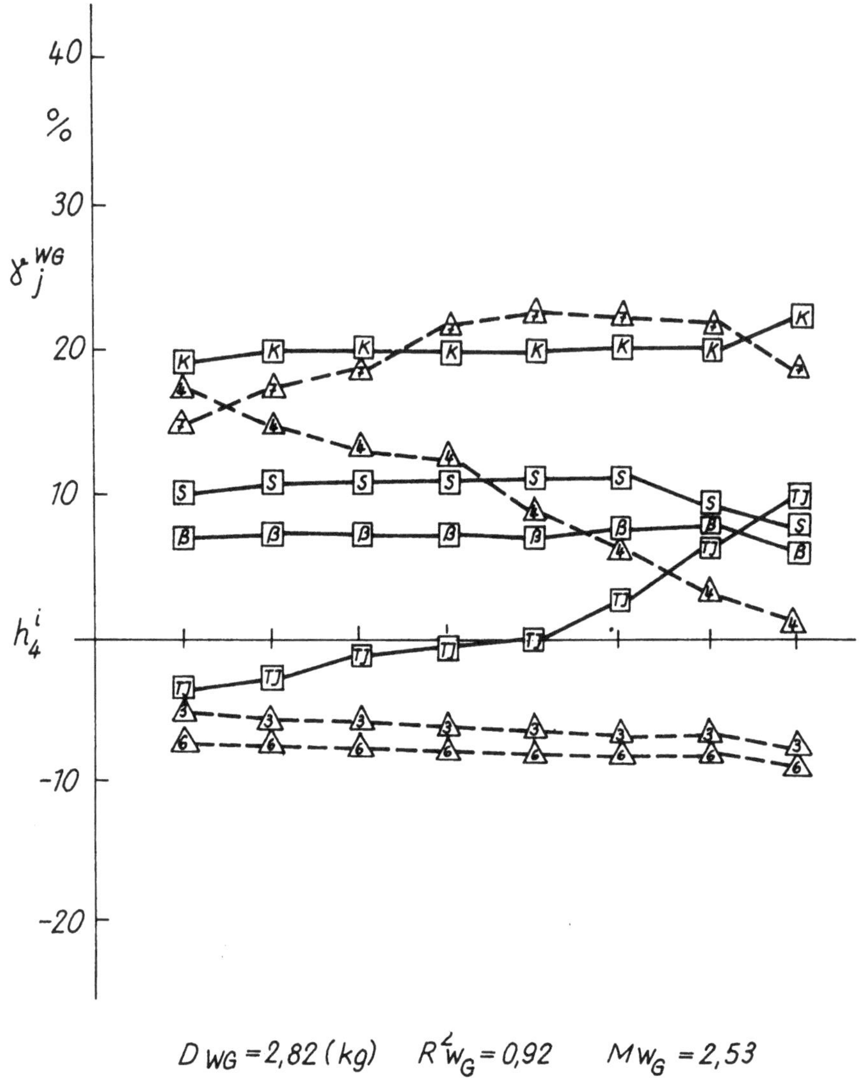

Fig. 43 (continued)

relation is explained in the following way: the short-living indirect trophic impact of the crab population on the hydroid SV6 is offset by the indirect continuous effect of consuming some of the insuforian population and, thus, the reduced intraspecific competition by juvenile nudibranches. Indeed, the analysis of the simulation results shows that the extent of the intraspecific competition among juvenile nudibranches abruptly increases in the beginning of the model and then falls to oscillate asymptotically below the mean values. This makes it possible for the entire predatory population to enjoy a smoother pattern of growth at higher values, and thus to insert an increased pressure on the hydroid population. The impact of the crab is, however, significant only within the $h_7^1 - h_7^3$ range, which is associated with the intensites $Z_7^F = (0.062 - 0.046\ kg/m^2)$. At lower Z_7^F the impact of the crab on the hydroid population is negligible. The negative impact of the set of trophic relations of mature nudibranches might be expected to be very high, but the impact of SV3 on hydroid in terms of γ_T^i and γ_3^i is seen, in Fig. 43, to be below that of indirect effects of the crab population. On the other hand, $\gamma_{T_T}^i$ and γ_4^i on the set of relations of juvenile nudibranches and infusoria, (SV4) shows a noticeable positive relation with the hydroids, which is probably again attributable to the effect of the intraspecific competition within nudibranch population.

The basic ecological patterns fit the research data qualitatively (Turpayeva, 1972), but the interdependence of vital functions is more involved and nonlinear, Figures 44 - 48. The dynamics of them are shown in detail for every computer experiment (only the most interesting situations are represented), where one single factor was varied, the remaining factors were invariable.

In Fig. 44 the variation of the mature nudibranch biomass is seen to have a significant effect on its juveniles only within the $h_T^1 - h_T^3$ range, or with the total biomass of mature nudibranches within the 0.013 - 0.019 kg/m^2 range (Table 27). As the effect of the biomass itself falls, the contribution of the nudibranch - hydroid SV3 set grows. For low biomasses of mature nudibranches, this can probably be attributed to the topic relations with the natural substrate and to the lower reproducibility of the population. Indeed, low biomasses of mature nudibranches are associated with small hydroid populations in the model or with periods of strong competition within the population. The dynamic study of the effect of the SV4 set reveals, Fig. 44, that the interval $h_4^6 - h_4^8$. Table 27, or the intensity range $Z_4^F = (0.069 - 0.046\ kg/m^2)$, is insignificant. The analysis of the effect on the mature part of nudibranch population, Fig. 45, demonstrates a strong positive effect of the SV3 set and of the degree to which young nudibranches have developed. As the magnitudes of the trophic "nudibranch - hydroid" relation (SV3) falls, the effect of the nudibranch population reproducibility increases. This ecological dependence of the dependence of the processes was noted by Turpayeva (1972), who found that a reduced number of juveniles results in a slower growth and that a negative impact on the mature part of the population results in slower reproduction. Consequently, the importance of compensation mechanisms increases.

The crab population in a fouling community may be said to exercise a regulative function. The polyphagous crab can participate in most sets

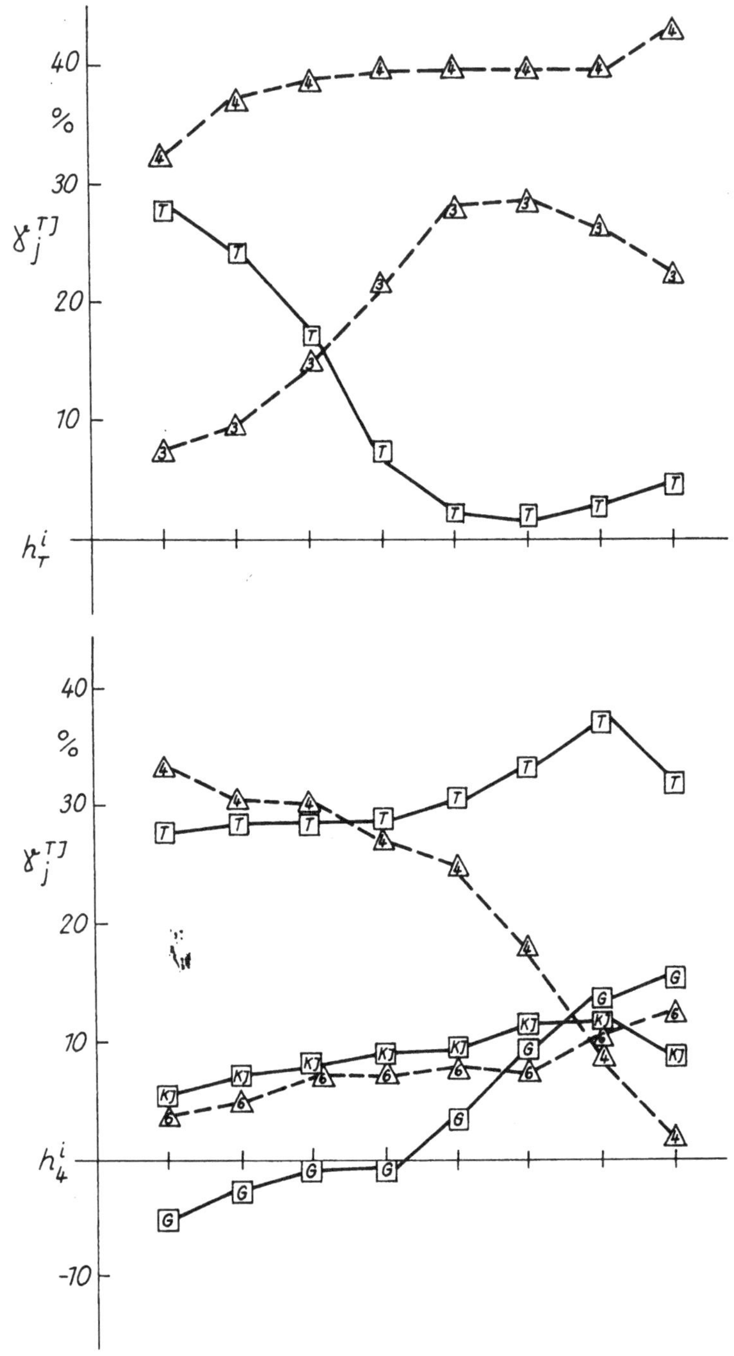

D_{TJ} = 0,003 kg R^2_{TJ} = 0,9 M_{TJ} = 0,002

Fig. 44

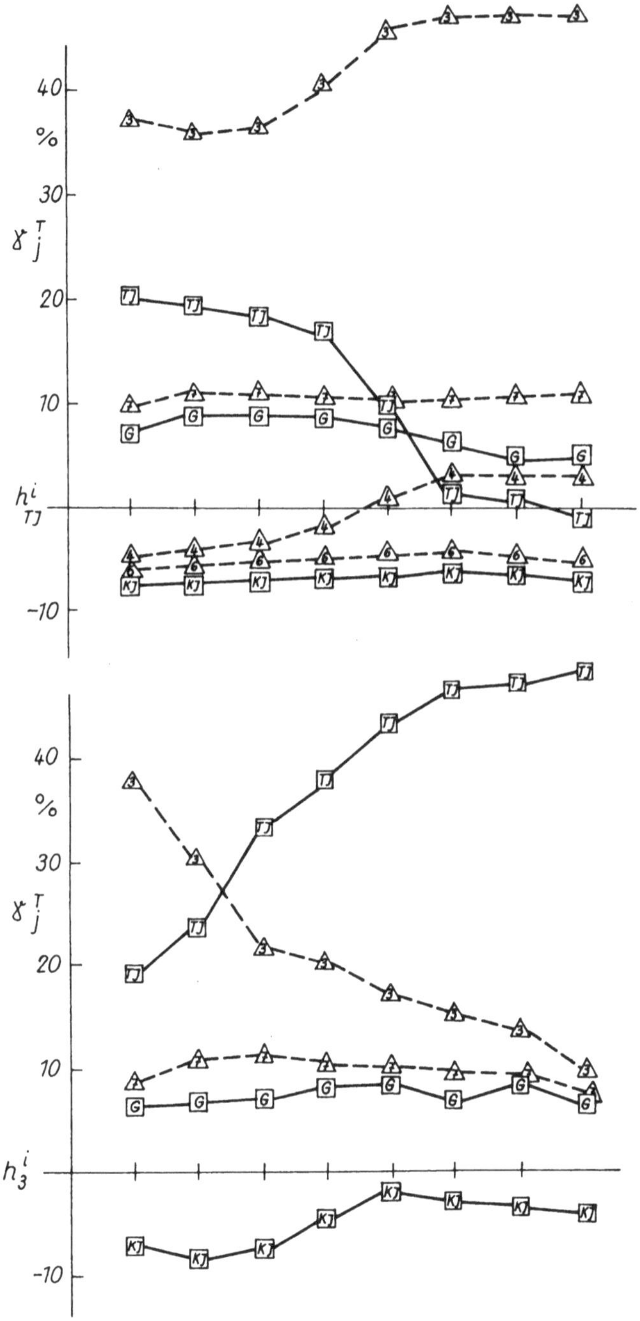

D_T = 0,003 kg R^2_T = 0,9 M_T = 0,0014

Fig. 45

of interspecific relations. The development of this population depends mainly on the availability of areas for the juveniles to settle (Gal'perin et al., 1981). Finally, adult crabs are cannibals. The extensive diet makes the crabs independent of any specific food, but complicates their relations with other populations. Periods of positive and negative impacts alternate, which is quantitatively confirmed by the data of the Figures 46 and 47. As the effect of the trophic "crab - nudibranch" set (SV6) on the juvenile crab biomass dynamics falls, that of the SV4 set grows. At highly intensive SV6s, i.e. within the $h_6^1 - h_6^3$ range or Z_6^F = (0.0040 - 0.0047 kg/m^2) range the "juvenile nudibranches - infusoria" relation has a negative effect on the dynamics of juvenile crabs, but as the significance of SV6 for juvenile crabs falls, the effect of the relation on the SV4 set becomes positive. The relations of juvenile crabs and the infusoria on which they feed influence the entire crab population. This effect is, however, maintained only within the small $h_7^1 - h_7^3$, or Z_7^F = (0.049 - 0.066 kg/m^2) range, outside which the dependence of juvenile crabs on the hydroid substrate is more important, Fig. 46. This is also true of the adult part of the crab population, Fig. 47. The expansion of the modelled infusorian community, which serves as the feed for the crab and nudibranch, depends mainly on the available area taken up by the hydroid population, Fig. 48. Juvenile nudibranches alone have an important impact on the dynamics of the infusoria biomass. The topic competition of the crabs and infusoria, which becomes acute when the hydroid biomass is in short supply, has a positive effect on infusoria whose potential to reproduce within a short time smooths the trophic shock experienced by this population of juvenile crabs.

In every modelled option, an increased salinity has a mildly negative impact. This can be attributed to an intensified expansion of predatory populations against the background of an insignificantly higher growth rate of the suppressed populations. The effect of zooplankton (the hydroid feed) concentration (β) on the response variance in a regression model is not large and neither are the γ_β^i . The calculation of γ_β^i disregarding the sets of trophic relations yielded 65 up to 70 per cent of R^2, and in the numerical results of the Tables 26 and 27 γ_β^i = 5 - 10 per cent. At first glance, the former values seem more plausible, for the total hydroid biomass falls with the concentration β. Our statistical processing method shows, however, the effect of the factors on the responses by variance.

In this context, the total deflections attributable in every statistical computation to the interspecific relations are more important in a regression model than the concentrations β in simulation experiments. Furthermore, in almost the entire model range of one simulation experiment the value of β was fixed and the coefficients γ_j^i were calculated

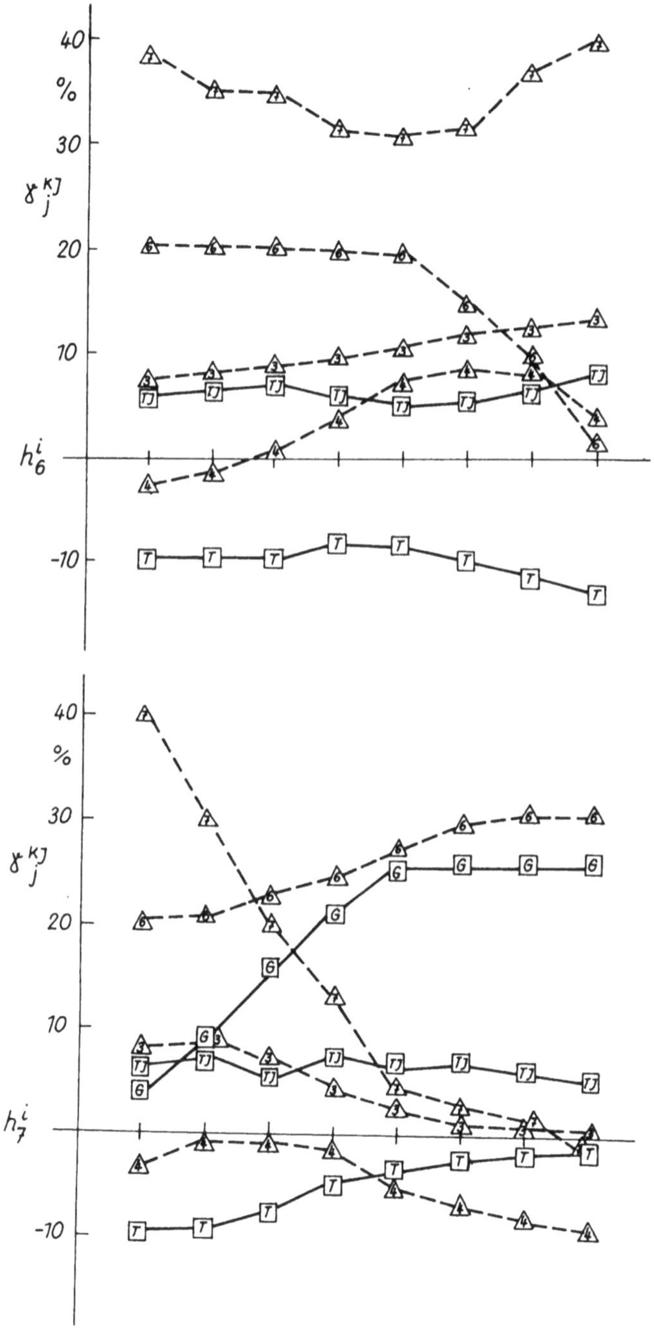

D_{KJ} = 0,003 kg R^2_{KJ} = 0,88 M_{KJ} = 0,002

Fig. 46

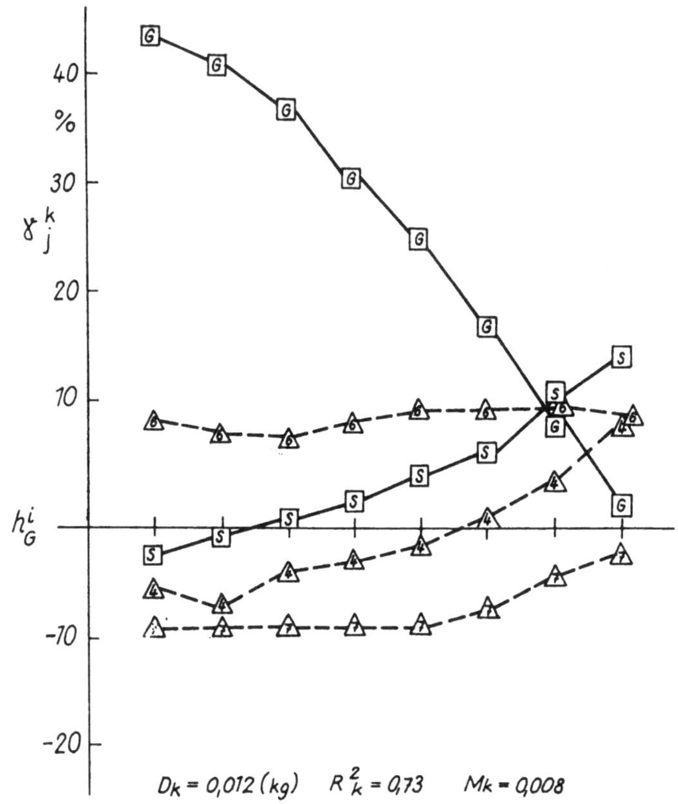

Fig. 47

from a matrix of data obtained in various simulation experiments. Therefore the variations of factors within one simulation experiment are more important for the variance of response functions.

Channels for the Biological Control of Bottom Coenoses

The qualitative evaluation of biological control channels and a analysis of numerical results suggest general conclusions true of other bottom communities which function in a similar way as well as the specific coenoses of this paper.

(1) The effect of indirect sets of interspecific relations on vital processes may prove more important than that of direct interactions. This is true, for instance, of the impact produced by the crab on the hydroid in a fouling community.

(2) The strength of the interspecific relations does not necessarily reflect the actual effect and so an additional analysis is needed to clarify the situation.

(3) Very important is that the maximum effect on the population of the dominant species in bottom communities over a short period seems less profound than long-term impacts of medium strength. Thus, dramatic expansion and death of the nudibranch population have been found to be more favourable for the hydroid population than the gradual expansion and small variations of the predator's biomass.

(4) The most efficient channels of biological control are impacts on the populations juveniles or attached populations at initial stages of development. Important variables at late stages are the sex - age structure of predatory populations and the state of interspecific relations. Turpayeva (1972) noted the importance of the latter variable but neglected the fact that this effect can be important over some time intervals if the suppressed population can trigger regeneration mechanisms and, what is more important, if a regulative population exists (like the crab population in a fouling community) to catalyze the intraspecific competition.

(5) Biological control by introduction of a new species may have an impact locally rather than over the entire time and space ranges. Furthermore, this local effect may be negative. Thus, in cases of numerous nudibranches migrating into the fouling community, it resulted in a

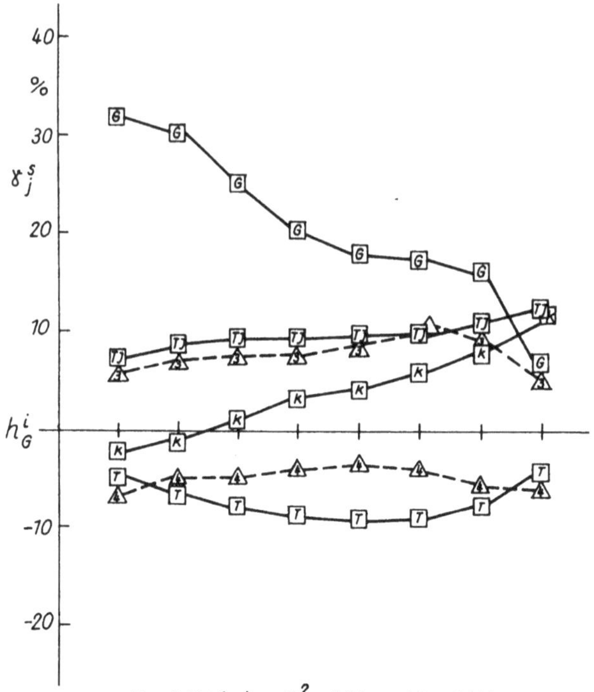

$D_S = 0{,}047\,(kg)$ $R_S^2 = 0{,}64$ $M_S = 0{,}028$

Fig. 48

quick suppression of the hydroid population and the subsequent death of almost the entire nudibranch population. Then the crab kept the nudibranch population on a very small size, while the hydroid population gradually recovered and its biomass eventually exceeded the average for similar environments.

(6) The proposed way to process the multi-dimensional statistics is useful in evaluating the channels for the biological control of bottom communities. Tested for fouling communities, the approach has been found applicable to a wider range of coenoses because its mathematical fundamentals are fairly general.

(7) For the first time, the quantitative estimates of the effect of the intensities of the interspecies relation over modelled discrete-value intervals may yield the strength of the impacts made by biological control and their sequences for obtaining the desired results either by a direct impact on the dominating species or, indirectly, by sets of interspecific biocoenotic relations.

4.2. Principles of Computer-Aided Event Simulation in Marine Ecology

Needs and Requirements

In recent years, the software costs have left the hardware costs far behind, especially in the modelling of complex systems. In the absence of a unified approach to modelling, the technological and programming costs of labour are inevitably excessive.

One way to reduce these costs is to create special classes of mathematical tools to solve problems in a standard formulation. Such tools are offered by theories of queueing systems, finite and probabilistic automata, games, etc. Dedicated simulation languages intended for specific ideologies of systems are used extensively. The most recent approach is the aggregative modelling by universal mathematical tools (aggregates), which are either piecewise-linear or piecewise-continuous (Buslenko et al., 1973).

Developed for non-biological systems, these tools fail to describe the variety of ecosystem functioning if the complexity of the biological processes increases. Even the widely used differential equations "impose" a qualitative behaviour on the biosystem and so simulation boils down to a manipulation of the variables. Similarly, in many cases the qualitative nature of the real process behaviour cannot be represented by the complex right-hand sides of the equations. Therefore, standard reduced models satisfying the need for unified software for ecology for some time were developed. The problems became then more complicated, but the results obtained with reduced models could not cope with the need to represent processes of higher complexity. Reusable models are even farther from materialization. Therefore it is natural that special emphasis is put on the development of block models whose modules can be used repeatedly in various models. These requirements make computer-aided simulation systems feasible in ecology.

First Results

Automation implies three basic components of the model (Buslenko, 1977; Gal'perin, 1986a & b):

- a standard structure of the data and the interface in the information retrieval system;
- standard mathematical description of element functioning and interaction;
- standard tools for the morphological description of the system (structuring, composition - decomposition).

Buslenko (1977) has embodied these priciples most comprehensively in an aggregative approach to simulation.

However as noted before (Chapter 1), the direct application of piecewise-linear aggregates (PLA) turned out to be ineffective for many reasons. Indeed, the motion of a PLA in a state polyhedron (Buslenko et al., 1973) is uniform and rectilinear and described as

$$\frac{dZ_\nu}{dt} = -\alpha_\nu \qquad (4.2.1)$$

where α_ν is a constant vector; ν is the basic state (an enumerable set); Z_ν is a vector of additional coordinates. By using (4.2.1), a time at which the aggregate jumps from the set of states Z into a new state can be found when the PLA reaches the polyhedron boundary. For the crab population (Chapter 2) one possible time of the "jump" is when various groups of individuals reach maturity. The only way to determine this time is to compute the individual's biomass dynamics by the well-known growth equation. Nevertheless, when the dynamics has been computed and the time determined, the next model interval cannot be studied by using equation (4.2.1) unless the time of a new "jump" has been computed. This is so because the PLA operates "by singular states" (Buslenko,1977) and the times of "jumps", when the boundary is reached, are assumed to be specified by the conditions of the problem, or given in advance.

The library of unified mathematical descriptions (models) of various ecosystem processes has also certain disadvantages such as (Gal'perin, 1983 a - e):

- incomplete ecological database and, consequently, possible "white spots" in libraries of unified models;
- increased needs in computer memory for storage of such libraries

and libraries of initial data and parameters;
- the need to develop algorithmical tools for integrating various process models into one simulation program.

For the fouling community, which consists of a few components, these problems could be overcome, mainly by compiling a standard morphological description of the ecosystem as a set of unified structural flowcharts (USF) of various components, (Section 1.4). USF's and mathematical descriptions of their elements and the interaction between them also led to standard algorithms for the generation of various model configurations and data processing:
- multi-level interfaces of subsystems and elements;
- multi-level flowcharts interfacing elements and subsystems.

Ecosystems in the Concept of Events Comparative Analysis of Principles

The unified structuring flowcharts of Section 1.4 consist of a finite number of elements. Because of its specific features, the system to be modelled is represented as a set of logical-heuristic algorithms implemented as software modules for this model rather than as a flowchart. The model element interfacing can be further improved. Furthermore, the description of the software for autonomous models and the model of the entire community make it obvious that the simulation software input/output and the transfer to simulation have been unified and, in the overall model, also the generation of an option by using the basic module of the entire simulation program. A major impediment to the replace the above mentioned model by a computer-aided simulation model* is the fact that the former is not built around a standard mathematical flowchart and the mathematical description is made of systems of differential, finite difference equations, and elementary function equations. Today there exist, to the best of my knowledge, only single-module CASMs which rely on standard mathematical tools such as piecewise-linear aggregates (Buslenko, 1977). This approach implies that the procedures processing quantitative characteristics have both benchmark times and "singular states" (Buslenko, 1973). For biological systems which are, in most cases, complex hierarchical and multi-component systems, such quantities for practical purposes cannot be specified in advance.

The model of every option has to be run "tentatively" in order to obtain such characteristics. This reduction of the system such as (3.2.1) - (3.2.36) to a standard form of a piecewise-linear aggregate is highly unwieldy and would require a tremendous programming effort.

* CASM

These difficulties can, in my opinion, be overcome by using unified mathematical flowcharts of biological processes similar to the model of the size of the crab population in Appendix 1. This model depends only on the numerical values of the process parameters, the composition and nature of values of the process parameters, the composition and nature nature of the mathematical description remaining invariable. Indeed the model can be adjusted to a new system without reprogramming.

Automation potential

A computer-aided simulation model requires the unification of the way in which the system servicing procedures as well as dedicated software procedures are connected to the basic set of programs. In reconsidering the results of Section 1.4 and Chapters 2 and 3 on the development of a simulation model by the proposed approach, adjustment maps, Sections 2.4 and 3.3, module and data libraries, standard print-out and input/output and unified mathematical tools were applied.

Consequently, computer-aided simulation models of biological systems may be obtained by the proposed approach, which is an application of the systems approach (Buslenko et al., 1973; Zeigler, 1976) to automation of the simulation of complex systems with the latest software tools. Such a model may yield quantitative results and make the methods of qualitative analysis applicable to simulation models in the framework of a qualitative theory of complex systems. The principle of standard form leads to tangible results in the qualitative theoretic research, which can be carried out along fairly general lines rather than for narrow ranges of systems.

In technology and economics discrete event simulation has been widely used, in particular, in the simulation of aggregated systems (Buslenko et al., 1973). Recall that in the aggregation approach an event occurs when the aggregate reaches a boundary (or the model time arrives at which the system state changes qualitatively) or an input signal is fed into the element. Quantitative changes are represented as aggregate motion in the state polyhedron while qualitative changes cause "jumps" into a new polyhedron (Buslenko, 1973; Kalashnikov and Nemchinov, 1983). For conventional ecological models the piecewise-continuous computer approximation of the system of equations is essentially a reduced aggregate. "Jumps" seem to occur continuously in such an aggregate, whereas a true jump caused by qualitative changes in the behaviour and the states of the process is treated as a transition of the approximating formulae. This is understandable because the computer realization of equation systems is a synchronous automaton.

Furthermore, in most ecological models, and in other models of complex systems, the "magic" of time is an inherent feature. Indeed, in most cases time as the basic variable makes it possible to arrange the events or activities in the system in a certain order. Time in itself is fairly rarely the cause of a change in characteristics. Thus, maturity is usually related to the individual's age. In reality this view is a sort of convolution, or aggregation of time-processes. Maturity should rather be considered the result of the interaction of numerous processes such as the cycle of matter, metabolism, etc. in the individual. It is only shortage of knowledge that is responsible for this convolution. Consequently, time is but an event ranking scale (ERS) On the other hand, quantitative changes of some variables may be regarded as a measure of real time because any quantitative change needs some measurable time.

With this understanding of time, the functioning of a complex ecosystem is a sequence of exogenous and endogenous events which result in a change of the system state. Even though this set of events is ranked on the model scale (which is most frequently time, but may be the area or type of the ground, coordinates, etc.), the difference between events culminating in quantitative or qualitative changes in the system is obvious.

Trivial as it seems, this discrete event concept of the functioning of an ecosystem enables researchers to classify some typical groups (or classes) of discrete events. The term "discrete event" is not used in this part of the article in any stringent sense like that of discrete modelling or the theory of probabilities.

Discrete Event Functioning of Marine Ecosystems

A comparative analysis of some generalizing papers on marine ecosystems (Kuznetsov, 1980; Patten, 1976; Paine and Levin, 1981) identifies four types or classes of events which are sufficient for an adequate description of the system functioning. No stringent typology or classification of ecological rules will be offered here, for this can be done with in the frame of the specific ecological concept used in specific research. As skills in event simulation improve, a generalized stringent typology will probably be developed. At this point we only want to show that the functioning of, for instance, a marine ecosystem can be discrete event-simulated, realized as an associated formal description, and this description may become the core of a computer-aided simulation model.

The event switching the system into a new state is either a change in the model ranking scale or some exogenous event. The event generates both quantitative and qualitative changes. Indeed, motion in time for some ecosystem components results in quantitative changes such as the increase or reduction of biogeneous elements in water, which, incidentally, may lead to qualitative changes such as the onset of the reproduction or migration of the population. A time change may result in qualitative changes of the system on the organism or population level. In this case, for the simulation interpretation of an event it would be necessary to keep track of prior quantitative and qualitative changes in the system, and this would entail excessive labour investments. Therefore, we think that, in case of complex systems such as ecosystems, the following hierarchy of events should be used without additionally testing whether a singular state has occurred or the polyhedron boundary has been reached. A change in the position on the scale leads to signal-events. The biological analogy of these events is, as noted above, for the marine bottom ecosystem the change of coordinates of the area or water volume, depth, type of ground or hydrodynamics. Events that are initiated by signalling ones may be referred to as compelling. Changes in water temperature, salinity, biogene concentration, etc. are of this type. Such changes bring about or, more precisely, make, the system change its state. Again, for some components this change is quantitative and for others qualitative. In addition to exogenous events, those resulting from internal or physiological changes of functioning should never be neglected. These events can be conventionally referred to as elementary because they occur in elements or subsystems of the lowest level in the structural flowcharts of the object to be modelled. Such events are, for instance, changes in the reproduction intensity or in the energy exchange among the individuals (for the organism level) or changes in the direction of the energy flow (for the population level). Finally, in this ecosystem, there occur interacting events such as changes due to interspecific relations, exhaustion of populations, and other maninduced events.

The Four Classes of Events Associated with Morphological Description

In the Figures of Section 1.4 for USF, signalling events are associated with changes in subsystems or physical-chemical conditions, or changes of interaction model types for communities. Compelling events are characteristics of the population feed or some of the physical-chemical conditions which directly influence the model of the community or trophic group. Elementary events occur for subsystems of the lowest hierarchy levels. Interacting events are naturally linked with subsystems of interspecific and inter-population relations and their models. A structured subsystem described in terms of discrete events may be incorporated into a higherorder description as a formal component and so methods of composition and decomposition can be applied (Kalashnikov, 1978). The discrete event description is sufficient for a model whose control and generation (not simulation) is automated. For illustration let us describe a generalized algorithm of the control module in such a simulation system.

A Generalized Algorithm of the Control Module in a Model Based on the Formal Description of Events

Step 1. Reading the adjustment card (or any other way to input the control data) of the simulation program, which may specify the number of biotic components to be modelled. One digit stands for the number of biological components the dynamics of whose basic states is simulated by the model; another digit represents the number of biotic elements whose state is needed in the model only as the input process (such as change in the biomass of the feed population in studying closed cycles in bottom communities). The number of abiotic factors in the card is equal to that of subsystems (and modules) in unified structural flowcharts. Auxiliary information may specify the duration of the model period, reference numbers of the necessary files and libraries, names of input modules, etc. Each ecosystem component is also associated with the numbers or types of unified flowcharts that are needed for its morphological description.

Step 2 - Step 3. Compiling and loading the common storage areas for control and auxiliary modules. Specification of the dimension, generation and primary distribution of external memory for the desired working files and input files.

Step 4. Input of the connection matrix for elements and subsystems of unified structural diagrams. This matrix embodies functional links in the system to be modelled. Furthermore, a matrix which ranks the events hierarchically is introduced to relate the four classes of events for every subsystem of elements according to the position in the rancing scale. The auxiliary margins store the number of interspecific relation modules, and list all the necessary modules. This matrix helps to adjust the modules for hierarchy computation and for setting the sequence of implementing and initiating the discrete descriptions.

Step 5. Checking the presence of the modules needed for simulation starts with a comparison of the tables of the generated module names for elements and subsystems of unified structural flowcharts and tables of module library headings. Later final tables of the names of existing modules and modules which have to be created for the current job are generated in magnetic units.

Step 6 - Step 7. The query about the data source for the desired modules whose names were not found in the library is fed to the console. The reply from the console updates the information for the creation of the desired modules. Generation of a full table of the names of the desired modules, then the creation of necessary procedures. Inputting auxiliary data on module connectivity and reference numbers of data sets. Activation of auxiliary modules and generation of reference "module data set" tables in order to make the modules consistent with the desired data sets and parameters of the mathematical tools.

Step 8. Computing the sequence for calling modules of signalling event count.

Step 9. Computing the sequence for calling modules of compelling event count.

Step 10. Computing the sequence for calling modules of elementary event count.

Step 11. Computing the sequence for calling the modules of interacting event count.

Step 12. To make the necessary amendments in the composition of modules and in the simulation software (including computer program texts), a reasonable description of the ecosystem is processed to generate a matrix of conditions for changing the functioning of elements. At subsequent intervals of the model time, this matrix is analyzed by a dedicated subroutine which determines the instances for amending the program modules. Simulation program amendments. Before the simulation starts, a file has to be created at this step for initial tests of the modules needed for this simulation step.

Step 13. In compliance with the computed sequences the text of a specific configuration of the simulation program is generated and written in a special file, while the data on this text are written in auxiliary storage areas. Print out, monitoring, and editing of the initial module texts. Editing of auxiliary information.

Step 14. Control transfer to the operating system and opening of data sets and libraries. Data input. Final adjustment of the simulation modules. Editing the text of the generated simulation program.

Step 15. Control transfer to the simulation control module.

Step 16. Conducting a simulation experiment. If the generation has to be controlled, the control is transferred to Step 4 after writing the data for correction of the simulation procedure.

Computer-Aided Simulation System (CASS)

Composition

Like any other simulation system, CASS has to include a database, a module (model) base, a data retrieval system, and generation-control-and simulation software tools. The CASS embodying the proposed approach would have original components. The data base should store libraries of parameters, tables associating USFs and event formalizing descriptions (block of structural transformations), and the parameters of appropriate mathematical equations and relations as well as data and results. If the above control and generation algorithm is employed, the base also includes standard event networks for standard USFs and, whenever necessary, indices of types of unified mathematical descriptions and algorithms. In addition to conventional libraries (Buslenko, 1977), the module base should store libraries of modules for interfacing and transformation, for scheduling first-level events (by the ranking scale) and lowest second-level events (with teir own calendars), modules of unified mathematical descriptions, modules for the conversion of system times of USF components (conversion of the event ranking scale, ERS), event count, and CASS generation and adjustment. Important is that the database should store, in particular, the numerical results of simulation experiments because the ecological data are often insufficient and so some simulation results have to be used in subsequent runs of simulation experiments as part of initial data. CASS should be equipped with tools for the qualitative and quantitative analysis of model structures as well as modelling results. Such tools are under development (Kalashnikov, 1978).

A Formalized Model of an Ecological Object in CASS

Using the event concept, a formalized model of the ecological object would be logically represented in a continuous by discrete form. In this representation discrete interactions are conventionally regarded as instantaneous changes in the system state or in the composition of processes, while continuous interactions are realizations of functional relationships among elements (Glushkov, 1975). Such relations are maintained on a finite time interval, while the start and end of their realization are regarded as acts of discrete interaction (Glushkov, 1975; Zeigler, 1976, 1984a & b). In our case, the number of discrete event classes associated with realization form changes in the system is univalently determined as four. These classes of events S, C, E, and I (signalling, compelling, elementary, and interacting) are combined to an algorithm, or a set of simple actions or statements. Unlike time modelling, we have an event ranking scale, g because the event scheduling statement PL is more sophisticated. For S this is PL_S (S, g) and for C, PL_C (C, (PL_S (S,q), Δt_c). Events of the type C are contingent on occurrence of the type S events and at time t_c, at which possible endogenous changes may occur in the associated system element (internal time). As a result, there arises a two-level event scheduler system with ramification by conditions. In a simple case the mark of an event is (C, PL_S (S, Δg)) and (S,g), while the scheduling leads to the generation of a new element from the ERS set (S,g + Δg) and (C,PL_S(S,g + Δg)).

If g is time or if it is not, but the internal time t_c of the element has no endogenous effect on the processes, the formalization is conventional. If g is not time and t_c is essential, then the event scheduler algorithms need synchronization. In essence, the effect of the time t_c is akin to the "jump" of an aggregate reaching the boundary. Discrete events influenced by t_c can be formally classified with the type E. It was noted above that a change in ERS can be regarded as a measure of real time. This is formalized by introducing a conversion statement $CN(g,T,t_c)$. This statement computes the instant of a total model time interval from a finite set T (SYSTEM CALENDAR) and the univalently associated time t_c of the element for which the statement $PL_c(C,(PL_s(S,\Delta g),\Delta t_c))$ is executed. Following this, as usually, the mark of an event with minimum value of time is accessed, t_c is entered into the variable T, and the algorithm C is executed. If the instant of an event of the C type is the same for several variables, then the conventional quasi-parallel approach can be used.

This formalization may seem redundant, but: a) computer-aided simulation needs redundant tools for the formalization of various situations; b) the ecological data on systems and their functioning are often ambiguous, which calls for a possibility to reduce the formalization techniques to one form; and c) time modelling is, in most cases, imposed on the ecologists. Thus, for bottom communities the most natural basic system variable would be the ground type or hydrodynamics (Kuznetsov, 1980), and for plankton communities, the depth (Vinogradov et al., 1973). For this reason these models are represented by very complex systems of partial differential equations. The introduction of ERSs and an ERS conversion statement in our approach may help to overcome these difficulties of considering the systems in a non-time scale.

Elementary and interacting events, the types E and I, are scheduled like those of the type C, namely

$$PL_E\left[E\left\{PL_c(C,(PL_s(S,\Delta g)),\Delta t_c),\Delta t_E\right\}\right]$$

and $PL_I\left[I,\left\{PL_E(\ldots),\Delta t_I\right\}\right]$.

This kind of scheduling is different from the conventional one (Glushkov, 1975) in that an event, say, of the type I, may occur not necessarily at certain instances but also before Δt_I ends if the chain of events which were triggered by a change of Δg in ERS leads to a new event of the type I. This signifies that a new element of the set SYSTEM CALENDAR $(I,T+\Delta t_I')$ is generated after $\Delta t_I' \leqslant \Delta t_I$. Consequently, the proposed formalized model performs two-level planning; firstly scheduling first-level events of qualitatively structural nature and, secondly, scheduling endogenous events resulting from changes of the states of the elements, or their functioning (second-level events). The set T (SYSTEM CALENDAR) exists for the purpose of reducing the event marks to a common denominator, and plays an auxiliary role.

In formalizing continuous discrete systems, a finite set of object types is introduced. Every object is a finite set of attribute variables. Objects of the same type are identically structured and differ in actual values of the attributes. Appropriate statements are introduced for the object generation. In the proposed approach the object generation statement GC(U), $U \in \{U_1, \ldots, U_n\}$ denotes, rather than the simple completion of the object type U with a new element, the inclusion of UDF into the overall structure of the system to be modelled and of the associated discrete event description into the "tree" of scheduled events. Conventional statements (Glushkov, 1975) of planning by condition, e.g. $PL_s(S,L)$ (where L is a logical function) and an assignment statement AC(a,Z), which assigns the value Z to the attribute, complete the formalized representation of the system. As in SIMULA-like modelling languages, processes and data conglomerates are used. Continuous processes are formalized in a conventional way (a Cauchy problem with check of the list of conditions for events to occur) and does not require many additional software tools to recognize the fact that the structure is specified in advance and event classes occur successively.

Numerical Analysis of an Ecological Example

As noted above, the proposed formalized model schedules discrete events at two levels. At the first level, events are structural-qualitative, and at the second level, they are contingent on whether functional relations between system elements are established or cancelled and on a change of the element state (reaching a certain level of the state value), which leads to qualitative changes in the system. The high complexity of event networks or chains in this formalization may require a preliminary numerical study aiming at reducing the size of event networks to be analyzed by the modelling algorithm. Second-level events are chiefly subject to this analysis because first level event networks depend on USF and ERS. The efficiency of the study can be tested for a simple and thoroughly studied ecosystem. Albeit not necessarily original, the numerical data on the dynamics of this system make it possible to assess the significance of event networks. The fouling community is a good proving ground for this approach.

Let us consider a simplified form of several standard event networks. ERS is dictated by signalling events (S) like changed coordinates of the substrate position in the sea, S_1 and time T when juvenile crabs or larvae of other populations immigrate, S_2. The compelling class of events (C) changes the fixed values of abiotic factors (superscript j) for various populations (suberscript i) such as water salinity and temperature $[C_1^{(i,j)}$ and $C_2^{(i,j)}]$ and intensity of hydrodynamics C_5. The same applies to the concentration of external food resources (superscript j) $[C_3^{(i,j)}$ and $C_4^{(i,j)}]$; and the intensity level of hydrodynamics C_5. The elementary class of events (E) changes the characteristic diet or the population feeding intensity $[E_1^{(i)}]$ and the response of compensatory mechanisms to inter-species interactions $E_2^{(i)}$. The interacting class of events (I) has only basic trophic and topic relations in the energy units (kcal). The trophic relations are: mature nudibranches - hydroids $[I_1^{(i,j)}]$; juvenile nudibranches - infusoria $[I_2^{(i,j)}]$; adult crabs - mature nudibranches $[I_3^{(i,j)}]$; juvenile crabs - infusoria $[I_4^{(i,j)}]$. The topic relations are: hydroids - juvenile nudibranches $[I_5^{(i,j)}]$; hydroids - adult crabs $[I_6^{(i,j)}]$; hydroids - juvenile crabs $[I_7^{(i,j)}]$; hydroids - infusoria $[I_8^{(i,j)}]$. Consequently, typical networks of first level events, (for USF, Section 1.4) may be represented in CASS as:

- for hydroids

$$S_1 \to \left\{C_1^{(i,j)} \cup C_2^{(i,j)}\right\} \cap \left\{C_4^{(i,j)}\right\} \to \left\{E_1^{(i,j)} \cup E_2^{(i,j)}\right\} \quad (4.2.2)$$
$$\to \left\{I_1^{(i,j)} \cap I_5^{(i,j)} \cap I_6^{(i,j)} \cap I_7^{(i,j)} \cap I_8^{(i,j)}\right\},$$

- for nudibranches

$$S_2 \to \left\{C_1^{(i,j)} \cap C_5\right\} \to \left\{E_1^{(i)}\right\} \to \left\{I_1^{(i,j)} \cap I_2^{(i,j)} \cap I_3^{(i,j)} \cap I_5^{(i,j)}\right\} \quad (4.2.3)$$

- for infusoria

$$S_1 \to \left\{C_\emptyset\right\} \to \left\{E_1^{(i)}\right\} \to \left\{I_2^{(i,j)} \cap I_4^{(i,j)} \cap I_7^{(i,j)}\right\} \quad (4.2.4)$$

- for crabs

$$\left\{S_1 \cap S_2\right\} \to \left\{C_5\right\} \to \left\{E_2^{(i)} \cup E_2^{(i)}\right\} \to \left\{I_3^{(i,j)} \cap I_4^{(i,j)} \cap I_6^{(i,j)}\right\} \quad (4.2.5)$$

The system (4.2.2) - (4.2.5) represents a standard event networks and a simplified morphological description of the ecosystem in terms of USF.

The statistical method shown above, Section 4.1, is used to investigate the significance of an event chain.

The numerical results of simulation experiments with the fouling community model were combined with data of field studies in proper parameter intervals into a matrix M. The matrix columns represented the values of assimilated population feed (biotic factors) and abiotic components for identical time intervals. Then the very large matrix (22,500 entries) was averaged over selected ranges in order to improve the statistical significance. Following this, the significance of standard events was numerically estimated statistically for computing the coefficients of the specific impact of the factor dynamics (i.e. the values of fixed levels of element states beyond which discrete events are expected to occur in CASS) on the variance of the value which characterizes the element state in UDF. The dynamics of γ is computed because useful data are eliminated from the initial matrix. Its variation shows if these data have an impact on R^2, Section 4.1.

The computation of such coefficients γ allows not only to determine insignificant standard event chains, or networks which can be neglected in simulation experiments, but also, whenever necessary, to identify the most important events of different types between which the system does not significantly change. This is very useful for the simulation in studies of system by "singular states" (as mentioned above in discussing the basic ideas of the aggregative approach) (Buslenko, 1977; Kalashnikov, Nemchinov, 1983).

To save space, only characteristic examples will be discussed. The Tables 28 and 29 present the values of the coefficients (as percentage of R^2) for levels of assimilated food for hydroids and nudibranches, the two dominating populations of the community. The columns are coefficients of the specific impact of possible events from a standard network (the system (4.2.2) - (4.2.5)) as a function of elementary events $E_1^{(i)}$ in this population and $E_1^{(j)}$ of a fixed level (in natural units) of UDF element states. In the tables 28 and 29 the multiple determination coefficient R^2 amounts to 88 - 98 per cent, which signifies that the unaccounted part of the variance $E_1^{(i)}$ is thus two to twelve per cent efficiency of the proposed method for estimating the significance of standard event networks in a formalized model.

Table 28-1

Statistical estimate of the specific effect of possible discrete events in a standard network (4.2.2) (hydroid population) when elementary events occur

Event type (I)			C_1	C_2	C_4	I_1	I_5	I_6	I_7	I_8	E_1
Units (II) / Element and fixing level (Elementary event-III)			SL salinity, 0/00	C^o, temperature	zooplankton β, thousands/m^3 of water	kcal	kcal	kcal	kcal	kcal	assimilated food, kcal
			1	2	3	4	5	6	7	8	9
			Specific effect, per cent of R^2 *)								
Assimilated food level in mature nudibranches kcal	1	0.00	0.00	0.00	43.60	2.07	-2.53 **)	0.02	-22.52	-7.63	1.51
	2	3.75	0.00	0.00	43.73	1.43	-2.64	0.02	-22.38	-8.11	439.1
	3	7.50	0.00	0.01	43.64	2.07	-2.42	0.02	-22.49	-7.65	576.6
	4	11.25	0.01	0.01	43.58	4.91	-1.55	0.03	-22.75	-5.58	390.7
	5	15.00	0.02	0.03	43.52	11.40	0.01	0.05	-22.21	-1.52	396.7
	6	18.75	0.02	0.03	43.66	10.72	0.00	0.04	-22.79	-1.93	626.8
	7	22.50	0.03	0.02	40.06	12.43	-0.02	0.04	-25.50	-2.30	513.6
	8	26.25	0.03	0.02	38.70	7.17	-2.35	0.05	-24.51	-6.88	463.0
	9	30.00	0.03	0.03	37.95	8.70	-2.27	0.06	-24.50	-6.58	461.4
	10	33.75	0.04	0.04	36.42	12.31	-2.24	1.14	-23.54	-6.31	537.7

*) The data in the row add up to R^2

**) The minus sign denotes negative correlation of the response as a function of the factor

Table 28-2

I		C_1	C_2	C_4	I_1	I_5	I_6	I_7	I_8	E_1
III	II	1	2	3	4	5	6	7	8	9
Assimilated food level of juvenile nudibranches, kcal 1	0.002	0.03	0.01	43.61	2.04	-2.53	0.16	-22.51	-7.63	1.51
2	0.98	0.04	0.01	41.83	1.34	-3.14	0.54	-22.54	-9.28	439.1
3	1.97	0.02	0.02	41.09	1.39	-3.77	0.65	-22.53	-9.44	576.6
4	2.95	-0.01	0.03	40.09	1.46	-4.30	0.60	-22.68	-9.21	390.7
5	3.93	-0.04	0.03	41.10	1.84	-4.65	0.46	-22.94	-8.24	396.7
6	4.92	-0.07	0.04	40.39	1.38	-5.27	0.58	-23.52	-8.82	626.8
7	5.90	-0.12	-0.02	40.96	1.76	-4.59	0.73	-24.87	-8.46	513.6
8	6.88	-0.11	-0.03	43.85	4.18	-2.39	0.31	-24.51	-4.60	463.0
9	7.86	-0.05	-0.05	46.12	4.33	-1.15	-0.11	-23.75	-3.24	461.4
10	8.85	-0.04	-0.06	46.69	4.25	-1.10	-0.29	-23.48	-3.00	537.7

Table 28-3

I			C_1	C_2	C_4	I_1	I_5	I_6	I_7	I_8	E_1
III		II	1	2	3	4	5	6	7	8	9
Assimilated food level in adult crabs, kcal	1	0.09	0.03	0.04	43.61	2.04	-2.53	0.16	-22.52	-7.63	1.51
	2	1.13	0.03	0.05	41.18	1.87	-2.78	0.81	-23.64	-7.95	439.1
	3	2.17	-0.04	0.08	43.07	2.26	-1.48	0.95	-25.89	-4.42	576.6
	4	3.20	-0.16	-1.21	42.98	-0.08	-2.62	0.59	-26.89	-5.37	390.7
	5	4.24	-0.19	-0.37	43.56	2.34	-3.19	0.48	-27.23	-4.51	396.7
	6	5.28	-0.20	1.81	46.97	2.72	-1.31	0.13	-26.87	0.32	626.8
	7	6.31	-0.21	2.42	45.66	3.85	0.26	-0.34	-24.60	3.80	513.6
	8	7.35	-0.23	3.17	38.56	6.26	4.14	-1.80	-14.28	11.98	463.0
	9	8.39	-0.24	3.09	29.51	9.70	8.17	-3.00	- 3.39	20.49	461.4
	10	9.42	-0.23	3.39	29.32	9.78	8.23	-3.01	- 3.26	20.63	537.7

Table 29-1

Statistical estimate of the effect of possible discrete events in a standard network (4.1.3) (mature nudibranches) for elementary events

Event type (I) / Unit (II) / Element and fixed level (elementary event) (III)			C_1 permille	C_5 m^3water/hr	$I_1 + I_5$ kcal	I_2 kcal	I_3 kcal	E_1 kcal, assim. food
			Effect in percentage of R^2					
			1	2	3	4	5	6
Assimilated food level in hydroid, kcal	1	0.00	-3.04	19.02	0.70	-55.62	-8.26	13.91
	2	68.57	-3.02	18.95	0.83	-55.65	-8.21	14.82
	3	137.31	-3.01	18.80	1.15	-55.67	-8.11	11.89
	4	205.97	-2.98	18.67	1.37	-55.71	-8.01	10.27
	5	274.62	-2.93	18.50	1.79	-55.70	-7.89	9.53
	6	343.28	-2.79	18.25	2.50	-55.59	-7.70	8.31
	7	411.94	-2.57	18.13	3.77	-54.81	-7.41	16.07
	8	480.59	-2.55	18.14	4.52	-53.98	-7.29	31.11
	9	549.25	-3.42	18.89	4.44	-53.73	-6.90	37.32
	10	617.90	-2.97	19.55	-0.02	-55.38	-8.67	30.64

Table 29-2

I			C_1	C_5	$I_1 + I_5$	I_2	I_3	E_1
III		II	1	2	3	4	5	6
Assimilated food level in juvenile nudibranches	1	0.02	-3.04	19.02	0.70	-55.62	-8.24	13.91
	2	0.98	-3.34	18.47	0.52	-55.25	-8.56	14.82
	3	1.97	-4.96	19.37	0.69	-51.86	-7.60	11.89
	4	2.95	-7.24	21.20	0.95	-47.40	-6.35	10.27
	5	3.93	-9.60	23.24	1.47	-42.89	-4.92	9.53
	6	4.92	-10.73	24.42	1.30	-40.62	-4.71	8.31
	7	5.89	-13.39	27.92	2.26	-35.79	-1.72	16.07
	8	6.88	-13.94	28.31	4.83	-28.20	3.44	31.11
	9	7.86	-14.49	28.90	4.88	-31.18	2.45	37.32
	10	8.85	-14.53	29.01	4.31	-30.17	2.37	30.64

Table 29-3

I		C_1	C_5	I_1 I_5	I_2	I_3	E_1
III	II	1	2	3	4	5	6
Salinity, permille 1	4.96 - 5.36	-3.04	19.02	0.70	-55.62	-13.24	13.91
2	5.76	-4.74	19.25	1.45	-51.86	-11.63	14.82
3	6.16	-6.06	19.15	2.56	-50.02	-10.27	11.89
4	6.56	-6.43	18.41	3.47	-50.27	-10.28	10.27
5	6.96	-6.68	17.60	3.92	-51.54	-9.85	9.53
6	7.36	-7.89	18.08	3.72	-51.50	-7.90	8.31
7	7.76	-9.65	18.71	3.39	-52.00	6.06	16.07
8	8.16	-12.19	17.89	8.88	-42.25	12.16	31.11
9	8.56	-15.50	18.27	10.29	-26.59	18.72	37.32
10	8.96	-16.59	18.71	12.68	-22.46	17.31	30.84

Table 28-4

I			C_1	C_2	C_4	I_1	I_5	I_6	I_7	I_8	E_1
III		II	1	2	3	4	5	6	7	8	9
Food concentration level for hydroid population, thousands/m^3 of water	1	0.40	0.03	16.81	43.61	2.04	-2.53	0.16	-22.52	-7.63	1.51
	2	5.36	-0.03	12.32	45.94	-1.15	-3.79	-0.35	-20.49	-10.29	7.59
	3	10.32	-0.07	11.08	47.56	-2.71	-4.31	-0.75	-18.81	-11.17	110.70
	4	15.28	-0.12	8.17	52.49	-2.88	-4.08	-1.33	-17.49	-10.23	134.72
	5	20.24	-0.18	6.12	60.08	1.67	-2.81	-1.90	-16.10	-7.43	461.40
	6	25.20	-0.25	3.01	68.59	0.77	-1.11	-2.47	-14.11	-3.32	477.60
	7	30.16	-0.22	1.94	69.56	2.92	1.04	-2.77	-12.89	0.94	489.40
	8	35.12	-0.23	-0.13	65.04	5.46	3.14	-2.50	-7.83	5.46	507.13
	9	40.08	-0.31	-1.63	65.93	7.75	3.72	-2.75	-3.56	7.14	653.70
	10	45.04	0.76	-2.07	50.76	8.89	3.49	-5.79	8.69	8.23	564.90

APPENDIX

UNIFIED MATHEMATICAL DESCRIPTION OF THE SIZE DYNAMICS OF THE CRAB POPULATION

The age group, representing the day of the emergence of specimens is denoted by τ. Since new arrivals join the population every day, τ coincides with the model time day T, the simulation step being one day. The original number (supplementing the population) of age groups at every instant of the model time is represented in the matrix A_τ. The functions describing the quantitative characteristics of the processes will be denoted by φ_j and the processes dictating the migration by ψ_l . The number of crabs $\hat{n}_\tau(t_\tau)$ of the same τ-th age group in the upper part of the fouling can be represented in the form

$$\hat{n}_\tau(t_\tau) = \sum_{l=1}^{P} \sum_{j=1}^{M} \tilde{a}_\tau \cdot \varphi_j(b_j^\varphi, c_j^\varphi, t_\tau) \cdot \psi_l(b_l^\psi, c_l^\psi, t_\tau), \quad (I)$$

where t_τ is the lifetime of crabs in the τ-th age group: $\tilde{a}_\tau$ is an element of the matrix of the initial population size; b and c are parameters of the functions φ_j and ψ_l ; l and j are subscripts of function types. For various kinds of number variations K not all biological processes are necessarily present for different intervals of the model time and specific existence conditions. This fact is taken into account by introducing a Boolean variable δ_k. Then equation (I) assumes the form

$$\hat{n}_\tau^K(t_\tau) = \sum_{l=1}^{P} \sum_{j=1}^{M} \tilde{a}_\tau \cdot \varphi_j(\delta_{jK}^\varphi(t_\tau), b_{jK}^\varphi, c_{jK}^\varphi, t_\tau) \times \quad (II)$$

$$\times \psi_l(\delta_{lK}^\psi(t_\tau), b_{lK}^\psi, c_{lK}^\psi, t_\tau),$$

where the subscript K implies that this term or parameter is true of the number of crabs in the τ-th age group whose variation may be classified with a certain type. The Boolean variable δ_K represents, in a multiplicative form, the functions φ_j and ψ_l . When φ_j or ψ_l assumes the value 0 (no process under way), in (II) assumes the value 0.

Writing the right-hand side of (II) as an algebraic sum the corresponding term would be helpful for further reasoning. For this purpose let us introduce

$$g_j^K = 1 - \varphi_j(\delta_{jK}^\varphi(t_\tau), b_{jK}^\varphi, c_{jK}^\varphi, t_\tau)$$

and $$m_l^K = 1 - \psi_l(\delta_{lK}^\varphi(t_\tau), b_{lK}^\psi, c_{lK}^\psi, t_\tau).$$

Then (II) is represented in the form

$$\hat{n}_{\tau}^{K}(t_{\tau}) = \sum_{l=1}^{P} \sum_{j=1}^{M} \tilde{a}_{\tau} (1 - (g_j^K + m_l^K) + g_j^K \cdot m_l^K). \quad (III)$$

The number $\bar{n}_{\tau}(t_{\tau})$ of crabs in the τ-th age group in the lower fouling layer can be represented, in the light of (II) and (III), as

$$\bar{n}_{\tau}(t_{\tau}) = \sum_{l=1}^{P} \sum_{j=1}^{M} \tilde{a}_{\tau} (1 - g_j^K) \cdot m_l^K \quad . \quad (IV)$$

The matrix of the membership of crabs of a given age to a group of given size will be denoted by $\sigma_{L\tau}$, where L is the current subscript of the matrix $\sigma_{L\tau}$ and can be computed by the formula

$$L = \alpha(t_{\tau}) \cdot d_S \quad , \quad (V)$$

where d_s is the column vector of sequential numbers of size groups and S is its dimension; $\alpha(t_{\tau})$ is the column vector defining, for a fixed time t_{τ}, the row of crabs of the τ-th age group of specified size. The resulting matrix D_{LT} of the numbers of crabs by the size in the current model year of joining the crab population is obtained by summation from the equation

$$D_{LT} = \sum_{\tau=1}^{T} (\hat{n}_{\tau}^{K}(t_{\tau}) + \bar{n}_{\tau}^{K}(t_{\tau})) \cdot \sigma_{LT} \quad . \quad (VI)$$

The final number of crabs in every size group is obtained by admitting for the crabs that survived the winter every size L that as in the population before the large-scale settling in the model of the current year. The number $\tilde{n}_L^F(T)$ of these "background" crabs for every size group is given by the equation

$$\tilde{n}_L^F(T) = \sum_{i=1}^{Z} \tilde{a}_{\tau} \cdot h_L^F(b_{iL}^h, T) \quad , \quad (VII)$$

where h_L^F is a function describing the elimination. With the matrix of background deflections denoted by F_{LT}, (VI) takes the form

$$\tilde{D}_{LT} = D_{LT} + F_{LT} \quad , \quad (VIII)$$

from which the population size is easily obtained from the formula

$$N_{\Sigma}(T) = \sum_{L} \tilde{D}_{LT} \quad . \quad (IX)$$

References

Afifi, A. and S. Eisen (1979). Statistical Analysis. A computer orientied approach.- New York: Academic Press.

Aizatullin, T. A. and A. V. Leonov (1975). Computer modelling the biochemical oxygen consumption in samples of sea water.- In: Chemico-oceanographic studies of seas and oceans. Moscow: Nauka, pp. 34 - 35 (in Russian).

Alekseev, V. V. (1973). The dynamics of the predator-prey system. In: A cybernetic approach to biology. Novosibirsk, Vol. 18, No. 5, pp. 922 - 926 (in Russian).

Alekseev, V. V. and M. S. Polyakova (1974). A model of vertical distribution of phyto- and zooplankton in the White Sea. In: Mathematical models of marine ecosystems. Kiev: Naukova dumka. pp. 5 - 7 (in Russian).

Bear, S. A. (1972). An algorithm of information-and-logical analysis.- Zhurnal obshshey biologii, Vol. 33, No. 3, pp. 359 - 372 (in Russian).

Beklemishev, V. N. (1951). On classification of biocoenotic (synphysiological relations.- Byulleten' MOIP, Vol. 56, No. 5, pp. 3 - 30 (in Russian).

Beklemishev, K. V. (1969). Regulation on the biocoenotic level of life organization.- Byulleten' MOIP, Vol. 74, No. 3, pp. 144 - 157 (in Russian).

Belyaev, V. I. (1978). Theory of complex geosystems.- Kiev: Naukova dumka (in Russian).

Belyaev, V. I. (1980). Mathematical modelling of ecosystems in seas and oceans.- In: Itogi nauki i tekhniki, Vol. 5, pp. 105 - 153 (in Russian).

Belyaev, V. I., A. I. Lenin, and T. S. Petipa (1974). Simulating the ecosystem of the pelagic part of the Black Sea.- In: Mathematical models of marine ecosystems. Kiev: Naukova dumka, pp. 10 - 13 (in Russian).

Birger, B. N. (1976). A generalized growth equation.- Ontogenez, Vol. 7, No. 6, pp. 645 - 646 (in Russian).

Boehm, B., J. Brown, H. Kaspar, M. Lipow, G. McLeod, and M. Merritt (1978). Characteristics of software quality.- Amsterdam, New York, Oxford: North-Holland.

Brand, S. (1970). Statistical and computational methods in data analysis.- Amsterdam, London: Elsevier.

Brauch, C. M. (1975). Intraspecific competition in Patella cochlear.- J. Anim. Ecol., Vol. 44, No. 1, pp. 263 - 281.

Brayer, M. (ed) (1975). Digital system design automation: languages, simulation of data base.- California Comp. Sci. Press, Inc.

Bukatova, I. L. (1979). Evolutionary modelling and its applications.- Moscow: Nauka (in Russian).
Buslenko, V. N. (1977). Computer-aided simulation of complex systems.- Moscow: Nauka (in Russian).
Buslenko, N. P. (1978). Modelling of complex systems.- Moscow: Nauka (in Russian).
Buslenko, N. P., V. V. Kalashnikov, and I. N. Kovalenko (1962). Lectures on theory of complex systems.- Moscow: Sovetskoe radio (in Russian).
Caughly, G. (1977). Analysis of vertebrate populations.- London: John Wiley and Sons.
Chugay, V. V., M. V. Gal'perin, and Ye. P. Turpayeva (1981). The energy flow through the crab ritropaneus population in the fouling community.- Okeanologiya, Vol. 21, No. 5, pp. 897 - 902 (in Russian).
Cunningham, G. (1978). A simulation model of primary production and carbon allocation in the creoosoteribish.- Ecology, Vol. 59, No. 1, pp. 37 - 52.
Cushing, D. H. (1953). Studies on plankton populations.- J. Cons. Perm. Int. Explor. Mer., Vol. 19, No. 1, pp. 3 - 22.
Cushing, D. H. (1959). The seasonal variation in oceanic production as problem in population dynamics.- J. Cons. Perm. Int. Explor. Mer., Vol. 24, No. 3, pp. 455 - 464.
Dame, R. F. (1976). Energy flow in an intertidal oyster population. Estuarine and Coast.- Marine Sci., Vol. 4, No. 3, pp. 243 - 253.
Davis, W. (1977). Operating systems. A systematic view. London, Amsterdam: Addison-Wesley.
Dijkstra, E. W. (1976). A discipline of programming.- Englewood Cliffs, N. J. : Prentice-Hall.
Dolgopol'skaya, M. A. and V. D. Brayko (1974). The fouling as an object in modelling of biological systems.- In: Research in structure and functioning mechanisms of marine ecosystems. Kiev: Naukova dumka, pp. 19 - 20 (in Russian).
Eberhard, L. and W. Hanson (1969). A simulation model for an arctic food chain.- Health physics, Vol. 17, No. 6, pp. 793 - 806.
Fleming, R. (1939). The control of diatom population by grazing.- J. Cons. Perm. Int. Explor., Vol. 14, No. 2, pp. 3 - 20.
Forrester, J. (1969). Urban dynamics.- London: The M.I.T. Press.
Gal'perin, M. V. (1983a). Simulation study of the fouling community: a model concept.- Okeanologiya, Vol. 23, No. 2, pp. 340 - 347 (in Russian).
Gal'perin, M. V. (1983b). Simulation study of the fouling community: mathematical formalization of the system of interspecific relations.- Okeanologiya, Vol. 23, No. 3, pp. 472 - 480 (in Russian).
Gal'perin, M. V. (1983c). Simulations study of the fouling community: computer implementation.- Okeanologiya, Vol. 23, No. 4, pp. 656 – 662 (in Russian).

Gal'perin, M. V. (1983d). Simulation study of the fouling community: a dynamic forecast of the system functioning modes.- Okeanologyia, Vol. 23, No. 5, pp. 876 - 881 (in Russian).
Gal'perin, M. V. (1983e). Simulation study of the fouling community: sensitivity to parameter variations.- Okeanologiya, Vol. 23, No. 6, pp. 1030 - 1038 (in Russian).
Gal'perin, M. V. (1986a). Principles of computer-aided simulation in biooceanology: the concept of the event approach.- Okeanologiya, Vol. 31, No. 2, pp. 317 - 321 (in Russian).
Gal'perin, M. V. (1986b). On two mathematical approaches to controlling biologically the evolution of bottom communities.- Papers on Mathematical Ecology, Budapest, Vol. 1, pp. 141 - 180.
Gal'perin, M. V. (1987). Principles of computer-aided event simulation in marine ecology. Ecological Modelling, Vol. 39, pp. 101 - 119.
Gal'perin, M. V. and V. P. Keondzyan (1981). Specifics of simulating biological communities.- Okeanologiya, Vol. 21, No. 4, pp. 719 - 725 (in Russian).
Gal'perin, M. V., V. V. Chugay, and Ye. P. Turpayeva (1981). Simulation of the crab population dynamics in the Sea of Azov fouling community.- Okeanologiya, Vol. 21, No. 3, pp. 545 - 554 (in Russian).
Gal'perin, M. V. and Ye. P. Turpayeva (1982). Simulation of processes in formation and functioning of the Sea of Azov fouling community.- Sevastopol, No. 5, pp. 98 - 101 (in Russian).
Gaure, G. F. (1933). A mathematical approach to struggle for survival.- Zoologicheskiy Zhurnal, Vol. 12, No. 3, pp. 170 - 177 (in Russian).
Gimel'farb, A. A., L. R. Ginzburg, R. A. Poluektov, Yu. A. Pykh, and V. A. Ratner (1974). Dynamic theory of biological populations.- Moscow: Nauka (in Russian).
Glushkov, V. M. (ed) (1975). Batch system modelling software. Kiev: Naukova dumka (in Russian).
Grahame,J. (1973). Breeding energetica of Littorina littorea (Gastropoda, Prosobranchiata).- J. Anim. Ecol., Vol. 42, No. 2, pp. 391 - 404.
Gurman, V. I. (1977). Singular problems in optimal control.- Moscow: Nauka (in Russian).
Gorstko, A. B. and F. A. Surkov (1976). On a mathematical simulation model of the bioeconomic complex in the Sea of Azov.- In: 2nd Workshop on Economic Issues in Nature Utilization. Moscow: Nauka, pp. 108 - 110.
Goodall, D. (1973). Ecosystem simulation in the US/IBP desert biome.- Proceedings of the 1973 Summer Computer Simulation Conference, Canada, Vol. 11, pp.777 - 780.
Halfon, E. (ed) (1979). Theoretical systems ecology.- New York: Academic Press.

Hannon, B. (1973). The structure of ecosystems.- J. Theor. Biol., Vol. 41, pp. 535 - 546.

Harris, G. (1972). A simulation model of productivity as a function of light intensity and water availability.- J. Ecology, Vol. 60, No. 1, pp. 19 - 40.

Holiny, C. S. (ed) (1978). Adaptive environmental assessment and management.- New York: John Wiley and Sons.

Holme, N. A. and A. D. MacIntire (1971). Methods of the study of marine benthos.- IBP Handbook, No. 16, Oxford.

Hughes, R. N. (1970). An energy budget for a tidal-flat population of the bivalve Scrobicularia plans (Da Costa).- J. Anim. Ecol., Vol. 29, No. 2, pp. 357 - 381.

Hughes, J. K. and J. I. Michtom (1977). A structured approach to programming.- Englewood Cliffs, N. J.: Prentice-Hall.

Ivlev, V. S. (1977). Experimental ecology of fish feeding.- Kiev: Naukova dumka (in Russian).

Jansson, B.-O. (1972). Ecosystem approach to the Baltic problem.- Bull. Ecological. Stockholm, Sweden: Research Committee/N.F.R., Vol. 16, pp. 1 - 82.

Jansson, B.-O. (1978). The Baltic - a system analysis and a semienclosed sea.- In: Charnoch and Eacon (eds.): Joint Oceanographic Assembly 1976, General Symposia. Stockholm: Plenum Press, pp. 131 - 183.

Jodan, E. (1975). Techniques of program structure and design.- Englewood Cliffs, N. J.: Prentice-Hall.

Kalashnikov, V. V. and B. V. Nemchinov (1978). Quantitative analysis of the behavior of complex systems by the method of trial functions.- Moscow: Nauka (in Russian).

Kalashnikov, V. V. and B. V. Nemchinov (1983). The organization of systems studies by aggregative simulation systems.- Math. Comput. Simulation, Vol. 25, pp. 502 - 512.

Kalman, R. E., P. L. Farb, and M. A. Arbib (1969). Topics in mathematical system theory.- New York: McGraw-Hill.

Khaylov, R. M. (1971). Ecological metabolism in the Sea.- Kiev: Naukova dumka (in Russian).

Kierstead, H. and L. Slobodkin (1953). The size of water masses containing plankton blooms.- J. Mar. Res., Vol. 12, No. 1, pp. 141 - 147.

Kogan, B. Ya. (1978). Computer-aided simulation and hybrid computers.- Voprosy kibernetiki, No. 46, pp. 5 - 20 (in Russian).

Kontorin, V. V. (1980). Mathematical modelling of the Baikal population.- Novosibirsk: Nauka (in Russian).

Krotov, V. F. and V. I. Gurman (1973). Methods and goals of optimal control.- Moscow: Nauka (in Russian).

Kurnetsov, A. P. (1980). Ecology of bottom communities in the World Ocean.- Moscow: Nauka (in Russian).

Kuenzler, E. (1961). Structure and energy flow of mussei population in a Georgia salt marsh.- Limnol. and Oceanogr., Vol. 6, No. 2, pp. 191 - 204.

Lenin, A. I. and V. V. Logvin (1977). Studying models of marine ecosystems by the DISTAL problem-oriented system.- Biologiya morya, Vol. 40, pp. 54 - 65 (in Russian).

Levin, S. A. (1974). Dispersion and population interactions.- Am. Mat., Vol. 108, pp. 207 - 228.

Levin, S. A. and A. Segel (1982). Models of the influence of predation on aspect diversity of prey population.- J. Math. Biol., Vol. 14, pp. 253 - 284.

Liepa, I. J. (1981). The dynamics of timber resources.- Riga: Zinatne (in Russian).

Lotka, A. J. (1925). Elements of physical biology.- Baltimore: Williams and Wilkins.

Lyapunov, A. A. (1971). On developing a mathematical model of input-output relations in the ecosystem of tropical oceanic waters.- In: The functioning of pelagis communities in tropical oceanic regions, Moscow: Nauka, pp. 13 - 24 (in Russian).

Lyapunov, A. A. (1974). A systems approach to studies of metabolic processes in biogeocoenosis.- Botanicheskiy zhurnal. Vol. 54, No. 8, pp. 1081 - 1092.

Margelef, R. (1968). Perspectives in ecological theory.- Chicago: Chicago Press.

Menshutkin, V. V. (1971). Mathematical modelling of marine populations and communities.- Leningrad: Nauka (in Russian).

Mesarovic, M. D. (ed) (1968). Systems theory and biology.- Proceedings of the III. Systems Symposium at Case Inst. of Technology, Berlin: Springer.

Madnick, S. E. and J. J. Donovan (1974). Operating systems.- New York: McGraw-Hill Book Comp.

Naumov, N. P. (1963). The ecology of animals.- Moscow: Vysshaya shkola, (in Russian).

Naylor, N. T. (1971). Computer simulation experiments with models of economic systems.- New York: John Wiley and Sons.

Nesis, K. N. (1977). General ecological concepts in application to marine communities. The community as a continuum.- In: Biological productivity of the ocean, series "Marine biology", Moscow: Nauka, pp. 5 - 14 (in Russian).

Neyman, A. A. (1963). On regularities in the composition of marine bottom biocoenoses. Zoologicheskiy zhurnal. Vol. 42, No. 4, pp. 618 - 621 (in Russian).

Nihoul, J. (1975). Modelling of marine systems.- Amsterdam: Elsevier.

Nikitin, V. N. (1947). The biology of ship fouling in the Black Sea.- Doklady AN SSSR, Vol. 8, No. 6, pp. 1183 - 1185 (in Russian).
Odum, E. P. (1971). Fundamentals of Ecology.- London: W. B. Saunders.
Odum, H. and E. Odum (1976). Energy basis for man and nature.- New York: McGraw Hill.
Paine, R. T. (1971). Energy flow in a natural population of the herbivorous gastropod Tegula funebralis.- Limnol. and Oceanogr., Vol. 16, No. 1, pp. 86 - 98.
Paine, R. T. and S. A. Levin (1981). Interidal landscapes: disturbance and the dynamics of pattern.- Ecol. Monograph, Vol. 51, No. 2, pp. 145 - 178.
Parker, R. (1968). Simulation of an aquatic ecosystem.- Biometrics, Vol. 24, No. 4, pp. 803 - 821.
Parker, R. (1973). Some problems associated with computer simulation of an ecological system.- In: Mathematical theory of the dynamics of biological populations. New York: Academic Press, pp. 269 - 288.
Partaly, Ye. M. (1974). The influence of the number of cruciforms in the plankton on the hydroid population in a fouling community.- Okeanologiya, Vol. 14, No. 4, pp. 719 - 723 (in Russian).
Patten, B. C. (1965). Community organization and energy relationships in plankton.- Oak Ridge Nat. Park, Tennessee.
Patten, B. C. (ed) (1972 - 1976). System analysis and simulation in ecology. New York: Academic Press, Vol. I - IV.
Pielou, E. (1977). Mathematical ecology. New York: John Wiley and Sons.
Platt, T., K. Mann, and R. Ulanowicz (1981). Mathematical models in biological oceanography.- Paris: The Unesco Press.
Poletayev, I. A. (1966). On mathematical models of elementary processes in biogeocoenosis.- Problemy kibernetiki, No. 16, pp. 171 - 190 (in Russian).
Radford, P. (1972). The simulation language as an aid for ecological modelling.- In: Mathematical models in ecology. Oxford: Univ. Press, pp. 277 - 295.
Razumov, A. S. (1969). Biological foulings in the potable and service water supply system and ways to combat them.- In: Biological foulings and ways to counter them. Moscow: Nauka, pp. 57 - 63 (in Russian).
Riley, G., H. Stommel, and D. Bumpus (1949). Quantitative ecology of the western North Atlantic.- Bull. Bingham Oc. Coll., Vol. 12, pp. 1 - 269.
Ross, J. (1970). Mathematical models of photosynthesis in a plant stand.- In: Prediction and measurement of photosynthetic productivity. Nageningen, Netherlands, pp. 29 - 45.
Shannon, R. (1975). Systems simulation. The art and science. Englewood Cliffs, N. J.: Prentice-Hall.

Shoberg, S. (1980). A mathematical and conceptual framework for models of pelagic ecosystems of the Baltic Sea.- Stockholm: Bull. ASKO laboratory, Vol. 27, pp. 3 - 105.
Shorygin, A. A. (1955). On biocoenoses.- Byulleten' MOIP, Vol. 60, No. 6, pp. 87 - 98 (in Russian).
Slobodkin, L. (1953). On social single species populations.- Ecology, Vol. 34, No. 2, pp. 430 - 434.
Smith, Y. M. (1974). Models in ecology.- Cambridge: Univ. Press.
Starostin, I. V. (1963). On the fouling of service pipelines in our southern seas and some techniques to combat this.- Proceedings of the Oceanology Institute, USSR Academy of Sciences, Vol. 70, pp. 101 - 123.
Starostin, I. V. and Ye. Ye. Permitin (1963). The species composition and quantitative development of the macrofouling on the sea bottom water pipeline of metal works (the Sea of Azov).- Proceedings of the Oceanology Institute, USSR Academy of Sciences, Vol. 70, pp. 124 - 141 (in Russian).
Steele, J. and D. Menzel (1962). Conditions for maximum primary productivity in the mixed layer.- Deep Sea Res. Oceanogr. Abstr., Vol. 9, pp. 39 - 49.
Sushchena, L. M. (1967). Substance and energy transformations in the amphibian population.- Doklady AN SSSR, Vol. 176, No. 3, pp. 703 - 706 (in Russian).
Sverdrup, H. (1953). On conditions for the vernal blooming of phytoplankton.- J. Cons., Perm. Int. Explor, Vol. 18, No. 3, pp. 287 - 295.
Svirezhev, Yu. M. and D. O. Logofet (1978). Stability of biological communities.- Moscow: Nauka (in Russian).
Straskraba, M. (1979). Natural control mechanisms in models of adequatic ecosystems.- Ecol. Model., Vol. 6, pp. 305 - 321.
Tamkevicene, Je. A. and I. A. Sestokas (1975). The impact of density on the growth of crayfish juveniles.- In: Fundamentals of bioproductivity of internal bodies of water in the Baltics. Vilnius: Vilnius Univ., pp. 337 - 338 (in Russian).
Timofeyev-Resovskiy, N. V. (1964). Some problems in radiation biogeocoenology.- Sverdlovsk (in Russian).
Turpayeva, Ye. P. (1967). On interspecific relations in fouling biocoenoses.- Proceedings of the Oceanology Institute, USSR Academy of Sciences, Vol. 85, pp. 43 - 48 (in Russian).
Turpayeva, Ye. P. (1969). Synphosiological (biocoenotic) relations in an oligomixt marine fouling biocoenoses.- Doklady AN SSSR, Vol. 89, No. 2, pp. 415 - 418 (in Russian).

Turpayeva, Ye. P. (1971). Man's impact on the ecosystems of marine bodies of water.- In: Biological productivity of the ocean and its utilization. Moscow: Nauka, pp. 264 - 282 (in Russian).

Turpayeva, Ye. P. (1972). The system of synphysiological relations in the marine fouling biocoenosis.- Proceedings of the All-Union Research Institute of Marine Fishery and Oceanography, Vol. 77, pp. 168 - 185 (in Russian).

Turpayeva, Ye. P. (1976). Experience in studying the relative rate of biocoenotic relations by mathematical experiment design.- In: Bottom fauna in the USSR's fringing seas. Moscow: Nauka, pp. 78 - 91 (in Russian).

Turpayeva, Ye. P. (1977). Fouling communities.- In: Biological productivity of the ocean, Series "Marine biology". Moscow: Nauka, pp. 155 - 162 (in Russian).

Turpayeva, Ye. P. and M. V. Gal'perin (1980a). The influence of settlement density of the energy budget of the cruciform Balanus improvisus in the Sea of Azov.- Okeanologiya, Vol. 20, No. 3, pp. 525 - 533 (in Russian).

Turpayeva, Ye. P. and M. V. Gal'perin (1980b). The influence of settlement density on the energy flow of the cruciform balanus improvisus in the Sea of Azov.- Okeanologiya, Vol. 20, No. 4, pp. 694 - 792 (in Russian).

Turpayeva, Ye. P. and M. V. Gal'perin (1982). A simulation model of the fouling community. 2nd Congress of Soviet Oceanologists. Sevastopol, Abstracts of Papers. Vol. 5, p. 133 (in Russian).

Turpayeva, Ye. P. and L. P. (1976). Numerical modelling of the population size of the nudibranch tenellia adspersa.- Okeanologiya, Vol. 16, No. 10, pp. 1437 - 1443 (in Russian).

Turpayeva, Ye. P. and R. G. Simkia (1975). The size and weight description, energy equivalent of the mass and the metabolism rate in mass species of the fouling biocoenosis in the Sea of Azov.- Zoologicheskiy zhurnal, Vol. 54, No. 8., pp. 1149 - 1153 (in Russian).

Turpayeva, Ye. P., V. V. Chugay, and V. I. Zatsepin (1980). The energy budget of the Rhitropanopeus Harrisidii tredentatus crab.- Zoologicheskiy zhurnal, Vol. 50, No. 1, pp. 32 - 42 (in Russian).

Turpayeva, Ye. P., M. V. Gal'perin, and V. V. Chugay (1981). Estimation costs in the developing population of Rhitropanopeus crab.- Okeanologiya, Vol. 21, No. 6, pp. 1084 - 1091 (in Russian).

Turpayeva, Ye. P., M. V. Gal'perin, and R. G. Simkina (1977). Food availability and energy flow in adult colonies of the perigoninus megas Kinne hydroid polyp.- Okeanologiya, Vol. 17, No. 6, pp. 1090 - 1101 (in Russian).

Turpayeva, Ye. P., V. P. Keondzhyan and R. G. Simkina (1976). Food availability and energy flow through a growing population of the hydroid polyp.- Okeanologiya, Vol. 16, No. 5, pp. 876 - 882 (in Russian).
Vinberg, G. G. (1962). The energy approach to studies of trophic relations and productivity of ecological systems.- Zoologicheskiy zhurnal, Vol. 41, No. 11, pp. 1618 - 1630 (in Russian).
Vinberg, G. G. (1976). The production of marine animals as a function of "growth" and "sustaining" productions.- Okeanologiya, Vol. 16, No. 5, pp. 854 - 867 (in Russian).
Vinberg, G. G. and S. I. Anisimov (1966). A mathematical model of an aquatic ecosystem.- In: Photosynthesizing high productivity systems. Moscow: Nauka, pp. 213 - 223 (in Russian).
Vinogradov, M. Ye. (1971). Research in the functioning of oceanic biological systems.- In: The functioning of pelagic communities in tropical regions of the ocean. Moscow: Nauka, pp. 5 - 12 (in Russian).
Vinogradov, M. Ye. and V. V. Menshutkin (1971). "Portraying" deterministic models of the pelagic ecosystems.- In: The biological productivity of the ocean, Series "Marine biology". Moscow: Nauka, pp. 261 - 276 (in Russian).
Vinogradov, M. Ye., V. F. Krapivin, V. V. Menshutkin, B. S. Fleishman, and E. A. Shushkina (1973). A mathematical model of the functioning of pelagic ecosystems in tropic regions of the ocean.- Okeanologiya, Vol. 13, No. 5, pp. 852 - 866 (in Russian).
Volterra, V. (1926). Variazione e fluttuazione del numero d'individui in specie animali conviventi.- Mem. Acad. Nazionale Linsei (ser. 6) N.2 pp. 31 - 113.
Vorovich, I. I., A. S. Goredov, A. B. Gorstko, Yu. A. Dobrovskiy, F. A. Surkov, and L. V. Epshteyn (1981). Intelligent utilization of the water resources in the Sea of Azov basin.- Moscow: Nauka (in Russian).
Watt, K. E. F. (1968). Ecology and resource management.- New York,London: McGraw-Hill.
Whittaker, R. (1975). Communities and ecosystems.- New York: McMillan.
Wilbur, H. M. (1976). Density-dependent aspects of metamorphosis in Ambistoma and Rara sylvatica.- Ecology, Vol. 57, No. 6, pp. 17 - 26.
Williamson, M. (1972). The analysis of biological populations.- London: Edward Arnold.
Wu, R. S. S., C. D. Levings, and D. J. Randall (1977). Differences in energy partition between crowded and uncrowded individual (Balanus glandula Darwin).- Can. J. Zool., Vol. 55, No. 4, pp. 29 - 41.
Wu, R. S. S. and C. D. Levings (1979). Energy flow and population dynamics of the barnacle Balanus glandula.- Marine Biol., Vol. 54, No. 1, pp. 83 - 89.

Zaika, V. Ye. (1975). Input - output equations of animal growth.- In: Quantitative aspects of organism growth. Moscow: Nauka, pp. 25 - 36 (in Russian).
Zeigler, B. P. (1976). Theory of modelling and simulation. New York: John Wiley and Sons.
Zeigler, B. P. (1979). Multilevel multiformalism modelling: an ecosystem example.- London: Academic Press, pp. 17 - 54.
Zeigler, B. P. (1984a). Multifaceted modelling and discrete event simulation.- London: Academic Press.
Zeigler, B. P. (1984b). System-theoretic representation of simulation models.- IEEE Transactions, Vol. 6, No. 1, pp. 19 - 34.

Bei Fragen zur Produktsicherheit wenden Sie sich bitte an:
If you have any questions regarding product safety,
please contact:

Walter de Gruyter GmbH
Genthiner Straße 13
10785 Berlin
productsafety@degruyterbrill.com